The Cosmic Circle

The Cosmic Circle

THE UNIFICATION OF MIND, MATTER, AND ENERGY

Robert Langs, M.D., and Anthony Badalamenti, Ph.D.

with Lenore Thomson, M.Div.

ALLIANCE PUBLISHING, INC.

ISBN 1-887110-04-6

Cover and interior design by Cynthia Dunne.

Alliance books are available at special discounts for bulk purchases
for sales and promotions, premiums, fund raising, or educational use.
For details, contact:

Alliance Publishing, Inc.
P.O. Box 080377
Brooklyn, New York 11208-0002

Distributed to the trade by National Book Network, Inc.
10 9 8 7 6 5 4 3 2 1

This book is dedicated to the Blumenthal Foundation, whose support gave us strength at times of crisis, and to whom we are forever grateful.

We are grateful to the Nathan S. Kline Institute for Psychiatric Research for providing facilities for the research studies on which this book is based.

Contents

INTRODUCTION ix

Chapter One Measuring the Unconscious? 1

Chapter Two Rethinking the Unconscious 7

Chapter Three Why Mathematics? 19

Chapter Four Initial Data 25

Chapter Five It's About Time 37

Chapter Six Natural Rhythm 45

Chapter Seven The Five Dimensions 57

Chapter Eight The Psychotherapeutic Dance 75

Chapter Nine The Skin of a Living Thought 79

Chapter Ten Effects or F/X? 91

Chapter Eleven Catching the Wave 107

Chapter Twelve Smashing Pumpkins 119

Chapter Thirteen The Law of Emotional Entropy 131

Chapter Fourteen The Physics of the Mind 145

Chapter Fifteen Working Hard or Hardly Working? . .155

Chapter Sixteen Poetry in Motion 163

Chapter Seventeen Caloric Entropy 171

Chapter Eighteen The Sounds of Silence 181

Chapter Nineteen Rethinking the Nature of the Mind . .189

Chapter Twenty The Cosmic Circle 199

REFERENCES INDEX 205

INDEX 209

Introduction

There is no greater mystery in the known universe, except the universe itself, than the human mind. We must view the mind as a special manifestation of matter. If this means invoking a property not yet included in the description of physics, this would not be the first time.

—Christian de Duve, *Vital Dust*

When we look at human bodies, what we normally notice is their surface features, which of course differ markedly. Meanwhile on the insides the spines that support these motley physiognomies are structurally very much alike. It is the same with human outlooks. Outwardly they differ, but inwardly it is as if an "invisible geometry" has everywhere been working to shape them to a single truth.

—Huston Smith, *Forgotten Truth: The Primordial Tradition*

It's been said that science is a limited mode of human cognition. For one thing, science demands a material object—something available to the senses and confirmable by others. The behaviors of that object must occur in time, so they can be measured and predicted. Finally, the resulting measurements must be mathematically expressible—that is, reducible to signs that are direct and unambiguous.

Clearly, there are many human experiences that cannot be explored in the terms I have just delineated. The "depth and breadth and height" of the soul's reach, of which Elizabeth

Barrett Browning speaks so eloquently, are scarcely quantifiable in a way that science would recognize as valid. Yet no one would argue that such dimensions of experience aren't real or significant. Human emotion would appear to slip through science's coarse grid like the sea through a fisherman's net.

But why should this be so? As Christian de Duve points out, the human mind is part of the natural world. Its products are demonstrably material, insofar as they expend energy. Our thoughts and our feelings occur as the brain consumes glucose and accomplishes neural connections.

Moreover, we know how we feel, in part, because of the way emotion affects our bodies. The heart beats faster, our eyes fill with tears, we laugh, our stomach muscles tighten. These physical aspects of emotion are part of our adaptational equipment as organisms. They move us to act in response to our external circumstances. We can see this most clearly, perhaps, in the sudden fear that makes us run before we know what we're trying to get away from.

No one would suggest that these physical aspects of emotion are tantamount to the feelings themselves. They are, however, the material framework on which our subjective experiences are constructed. This is what science can measure and predict. Not feelings themselves, but their process of occurrence—their relationship to an external stimulus.

The knowledge science has granted us of such things has not abrogated the mysteries of individual perception. Rather, it's shown us the unity underlying human variation. We know, for example, that the physical signs of emotion are definitive and universal; they transcend culture. All people rely on them to interpret the intentions and responses of others.

But emotional responses are not confined to their biological effects. They are reflected in human products of all sorts—art, religion, music, literature. Psychological research has attempted to determine the ways in which particular artistic endeavors are fueled by emotional conflict, but these efforts have met with little success. It is possible, however, to approach this area of research from a different angle.

Our creative productions are all forms of human expression. As such, they share common ground with the physical aspects of emotion. That is, they function as a form of communication. Like the physical aspects of emotion, which are largely involuntary, our communicative productions reflect both our conscious and unconscious responses to external experience. We can see this most clearly in our use of language. And therein lies our tale.

As biologists have long pointed out, language is unique among our capacities for biological communication. Words are not definitive—in the way, for example, a galvanic skin response to a stimulus is definitive. Words are ambiguous. They can mean several things at the same time. Our brain is actually programmed for linguistic ambiguity—for the ability to say one thing and mean another at the same time.

As a psychoanalyst, I have become convinced that linguistic ambiguity evolved specifically as a vehicle of emotional adaptation, which sometimes parallels, but may also subvert and contradict our behavioral responses to emotion. My way of practicing therapy is predicated on the ways in which clients' stories often function at two levels of meaning.

I am not speaking here about the classic Freudian distinction between, say, the manifest text of a dream, and its hidden psychological meaning, which is held to indicate unresolved childhood conflicts. I am actually extending Freud's insight into the domain of ordinary conversation, and I'm relating the process to the client's unconscious attempts to adapt to his or her present context.

The client's words have a direct, surface meaning, which reflects something he or she is thinking about consciously; but they also reflect, indirectly, an unconscious attempt to adapt to the environment emotionally. This second level of meaning is apparent *only* when a client tells a story—that is, when a client uses narrative imagery that can encode unconscious emotionally charged material. Explanations and intellectualizations have only a single, direct meaning, and don't appear to be adaptive in the same way.

My working hypothesis has been that, like the physical concomitants of emotion, communicative adaptation by way of storytelling is universal. Its output follows certain laws, happens in a predictable way—no matter what language we are using. Hence, communicative adaptation can be linked scientifically with other forms of measurable behavior.

In this respect, the therapeutic enterprise can serve as a kind of laboratory. It is entirely possible, for example, to quantify the number of stories and intellectualizations that occur in a therapy session. All manner of communicative elements—Who is speaking? For how long? Positive or negative imagery? New topic?—can be counted and measured as objective phenomena. The very idea that such things might be temporally regular, predictable, and scientifically expressible was enough to motivate research in this direction—not only because the findings could determine a more accurate way of defining and treating psychological problems, but because it would tell us something important about human relationships and the human mind.

All people tell each other stories. We move in and out of narrative expression constantly, communicating, in the process, our unconscious attempts to adapt to our circumstances—and responding to similar attempts by our speaking partner. Why shouldn't this tendency follow the same kind of laws as other natural processes? In my search for a way to observe and measure this tendency, I was fortunate to meet Anthony Badalamenti, Ph.D., a mathematician and systems theorist who worked with me to develop a protocol and to model the results mathematically.

I had the psychoanalytic data and the communicative theory, and Tony had a mind for the kinds of mathematical models that could unearth the "invisible geometry" underlying a deceptively wide variety of communication styles. We began the research with a study of recorded psychotherapy consultations. The results we found, however, encouraged us to extend the work to couples in dialogue and even to emotionally relevant monologues.

We also explored the way in which poets, writers, and speakers *use individual words.* The lawfulness of word selection was astonishingly consistent—whether we were dealing with the plays of Shakespeare or the signs of chimpanzees! What began as a study of communication and adaptation in psychotherapy has ended, at least for now, as an investigation of the fundamental properties of all types of human (and protohuman) communication.

I do not wish to suggest that research of this sort can somehow capture the human heart and reduce its complexity to matters of physics. I want to emphasize again that we were not attempting to measure emotion or meaning as such. We measured the ways in which humans *convey* emotional meaning—the output of emotion, its *vehicles.* Moreover, our quest was, quite frankly, as romantic as it was intellectual. We were interested in the position we have, as humans, in the natural universe, our place in the unchanging laws of temporal existence. The potential of such research is spiritual as well as scientific.

Indeed, the potential of this research extends in many directions. Our work can illuminate certain aspects of the therapeutic process in a way that portends a formal science. But it also suggests how effective communication takes place in everyday life as well. It is our fervent hope that the ways of understanding ourselves and others revealed through our work will offer some insight into the fundamental unity that underlies our surface differences—a unity that we share with nature at large.

—ROBERT LANGS, M.D.

O Nature, and O soul of man! how far beyond all utterances are your linked analogies! not the smallest atom stirs or lives in matter, but has its cunning duplicate in mind.

—THOMAS MELVILLE

CHAPTER ONE

Measuring the Unconscious?

The concept of the unconscious posits nothing;
it designates only my unknowing.

—C. G. Jung to Pastor Max Frischknecth

ALTHOUGH FREUD REGARDED his work as a form of science, because it was based on observation and the formulation of theory, psychoanalysis has never been what is usually called a "hard science"—that is, one based on data that are readily quantified and verifiable. After all, how can one measure a client's thoughts, feelings, transferences, resistances, dreams, and fantasies? Such things are, by nature, subjective and intangible. Accordingly, Freud complained:

> *I always envy the physicists and the mathematicians who can stand on firm ground. I hover, so to speak, in the air. Mental events seem to be immeasurable and probably will always be so.*

It has proven difficult to dispute Freud on this issue. The marriage of psychoanalysis and science has been a rocky one at best. Analysts who have attempted to join the two realms have ended up with little more than statistical correlations—evidence that one measurable event is related to another. For example, statistics indicate that patients who like their therapists report better therapeutic outcomes than patients who don't. It's tempting to leap from this statistical finding to a causal assumption—the idea that liking one's therapist ensures therapeutic success. But the events may be correlated simply because patients who like their therapists idealize their efforts. Useful as they are, statistics can't answer questions about how things actually work.

Indeed, the therapeutic enterprise succeeds and fails regularly, and there are no adequate explanations for either result. The very fact that three hundred forms of treatment exist, many of which conflict with each other theoretically and technically, virtually defines psychotherapy as a healing intervention without scientific foundation. Like a bird without feet, its flight has been breathtaking, but ungrounded.

Even in an everyday sense, a genuine science of the emotional domain would benefit us. We have managed to generate technologies whose consequences are increasingly beyond our control. Not least among these consequences is the undermining of our faith in the human ability to influence what happens to us. Our social context is complex and rewarding, but the existence of multiple perspectives in a rapidly changing world creates uncertainty and moral ambiguity.

Given our ability to forge a group of elaborate material sciences, we should be able to develop a complementary science of the emotional mind and its adaptational principles. Interest in emotional concerns and conflicts has never been higher. Self-help books are legion. But commonsense observations and neighborly advice about such things as finding love and maintaining it are the domain's equivalent of bloodletting and purging in nineteenth-century medicine. Until we discovered germs and viruses and understood their functional contribution to disease, our medical technologies were naive at best, primitive at worst.

We need, as it were, to invent an emotional microscope. Unless we really know how the mind adapts to emotionally charged experiences, our attempts to facilitate or remedy the process might easily do more harm than good.

PROBLEMS IN CREATING SCIENCE

Why has a formal science of psychoanalysis seemed like an impossible dream? The answer to this question takes us back to Freud's idea that psychoanalytic data—its observables—consist of inherently immeasurable entities: thoughts, feelings, and the like. Freud was also laboring under nineteenth-century ideas about scientific objectivity. As far as he was concerned, the client's inner experiences had no predictable connection to the interaction with the therapist. The therapist was merely observing and interpreting.

These are unnecessarily limiting assumptions. The word *datum* simply means "given"—the substance of what can be observed. Freud's point, of course, was that the givens of psychoanalysis are mental events, part of a person's experience—hence immaterial. They can't be observed or quantified. Science requires that we measure the evolution of events and entities over time. But to believe that *no* mental product or entity can be measured over time is to create a philosophical cul-de-sac that is unwarranted.

At bottom, the client's subjective state is *not* the basic datum of psychoanalysis. A therapeutic exchange is not an internal monologue; it's a *dialogue* that takes place with another person. That dialogue consists of words, which are the fundamental givens of a session. The client's thoughts and feelings are the *content and meaning* of those words.

Freud's preoccupation with meaning led him to undervalue the context in which meaning occurs. He wasn't thinking about the vehicles that carry meaning—words and sentences in particular, and he wasn't thinking about his own influence on those vehicles. As a result, he understood his patients' words to reflect their respective internalized heritages and missed their adaptive aspect—as responses to himself and the event of being in therapy.

Although a science of subjective thoughts and feelings is all but impossible, a science of human, language-based communication is a very real possibility. Communication is as much an output of the mind as electrical activity is an output of the brain, and bile an output of the liver. Measuring these outputs can tell us a great deal about the organs and organ systems creating them.

Any effort to create a psychoanalytic science must entail a shift of perspective. We need to stop thinking only about what is happening in the mind of the client (or therapist) and start to think more about what is happening between two people—their interaction and dialogue. Freud was not centered on the data of the therapeutic exchange. He was focused on the meanings he inferred from his clients' words, which he believed revealed their inner conflicts. In his view, those conflicts were the result of having repressed an emotionally charged experience that could not register consciously.

This is one reason Freud overlooked his clients' productions as attempts to adapt to himself and the therapeutic experience. His idea was that a client's life problems were caused by the ongoing attempt to keep repressed conflicts at bay. His aim was to induce the client to bring the repressed information back into awareness and to submit it to rational understanding. With few exceptions, all dynamic therapies are guided by this particular paradigm. As psychoanalyst Theodor Reik (Saturday Review, January 11, 1958) nicely put it:

The repressed memory is like a noisy intruder being thrown out of the concert hall. You can throw him out, but he will bang on the door and continue to disturb the concert. The analyst opens the door and says, "If you promise to behave yourself, you can come back in."

Most therapeutic techniques are based on this idea—that unconscious "intruders" will ultimately become apparent in the ongoing relationship with the therapist. This is what psychoanalysts generally call "transference." Old coping strategies, patched together in childhood to deal with long-forgotten traumas, reassert themselves and interfere with healthy intimacy in our current relationships. The therapist's job is to help the client become aware of these unconscious coping mechanisms—to talk about them rather than to "act them out," and hence to get them under conscious control.

This particular conception of the "unconscious," however, has kept psychoanalysts from attempting the kinds of measurements that characterize a legitimate science. It has moved psychoanalysis quite nearly into the domain of parapsychology. Therapists can seem more like "ghostbusters" than healers.

In fact, it's worth looking at the odd parallel between psychology and parapsychology. In the realm of parapsychology, ghosts haunt houses somewhat unwittingly. Their last moments were traumatic and they don't realize yet that they're dead. Trapped between matter and spirit, the ghosts try to cope by repeating the behaviors that led to their death, and some of their actions have an effect on people who are still alive. The parapsychologist's role is to get in touch with the ghost and explain what happened so the entity can let go and move on.

Somewhat similarly, a therapist assumes that the ghost of a client's past trauma is trapped between conscious and unconscious realms. In an attempt to cope, the client has been unwittingly repeating behaviors that led to the trauma, and the result has been problematic relationships. The therapist's role is to help the client get in touch with the ghost so that old defense patterns can be laid to rest.

Although the image has had enormous value, and I would not minimize the help and guidance that therapists and counselors provide, it is probably no coincidence that parapsychologists have the same problem that Freud had in measuring the phenomena of their profession. Like spirits, past traumas are notoriously difficult to quantify and predict. Moreover, Freud understood himself as a go-between, helping to enlarge the client's personality by admitting into consciousness unwanted aspects of the psychic makeup. How could he possibly quantify the impressions, inferences, and inner dialogues that constituted his manner of playing that role?

Whatever the latter-day changes in psychoanalytic thinking, the ideas about psychoanalytic content remain very much the same. Even when the therapeutic interaction is acknowledged, as it is in the systems-therapy approach, the focus remains on the client's inner experience of that interaction rather than on the dialogue and relationship with the therapist. The communicative dialogue between client and therapist goes unaccessed as a rich source of measurable material.

The research described in this book became possible only by extending this meaning-centered paradigm and, in some cases, by departing from it. We had to stop thinking about a client's mental process and concentrate instead on the communicative interaction between two people. We had to abandon the idea of observer and observed and needed to focus instead on mutual adaptation through dialogue. This was the only way to move from impressions and statistics to something approaching a formal science of the emotional mind.

CHAPTER TWO

Rethinking the Unconscious

What we need most, is not so much to realize the ideal as to idealize the real.

—H. F. HEDGE

AS SUGGESTED IN the last chapter, Freud's theory was centered on the nature and causes of psychic pain and the means of avoiding it. His idea was that when some primal tendency, sexual or aggressive, cannot express itself in its original form, it finds expression as a symptom. Although this idea was, in fact, derived from the physical sciences, Freud's focus kept him from seeing some of its more important implications.

Repression, in his view, was a kind of psychic boundary that keeps certain ideas and their emotional content from conscious experience—much as the skin of a balloon keeps air from escaping into the atmosphere. His interest was concentrated, however, on the inner surface of that boundary—the psyche's refusal to let inner (instinctual) forces into conscious awareness. Thus, Freud missed the fact that a boundary is inevitably two-sided. It has a relationship to and an interaction with its external environment.

Think about the skin of an apple. It protects the inside of the apple and keeps it distinct from the environment. But it also permits contact with that environment, allows for an exchange of elements. Our own skin is a boundary. It protects our system from certain environmental forces, but it is also part of the environment. It admits thermal energy and emits heat and water. One can think of the mind in the same way. It has contact with its immediate environment, and its outputs—in language and communication—reflect the nature of that contact.

Every natural system has a boundary, which keeps it distinct from its external environment, but is also the site of exchange between the system and the outside world. One of the great accomplishments of physics has been the recognition that activity on the boundary of a system can tell us what is happening and not happening within the system.

For example, the net sum of electrical charges within a volume is equal to the electrical flux passing through its surface. This means that by measuring the action taking place on a volume's outside, we can make estimates about what's going on inside. Mathematicians have worked in conjunction with physicists to determine exact methods of calculating such relationships between the boundary and interior of a system. This book is about applying the same mathematical models to psychological boundaries, as manifested in the use of language.

ADAPTATION TO THE ENVIRONMENT

We are constantly adapting to our context, at many levels of experience. We simply aren't conscious of all the perceptions we are registering. Think about daydreaming in heavy traffic. The light changes, a man two cars back is honking his horn. Consciously, we adapt by hitting the gas and moving the car. If we had to think about how to do this—how to increase our adrenaline and manipulate the muscles in our pedal foot—we would not get very far.

Consciously, we are aware of the surface perceptions that move us to respond. Unconsciously, however, we are registering and responding to a whole network of perceptions that we don't process consciously. There is nothing mysterious about this process. Our sensorium is structured to screen out things as well as to focus on them.

The same thing happens with perceptions that trigger our emotions. Perhaps we feel angry because the man who honked his horn forced us to attention. This is not an earth-shattering impression, but it's irrelevant to our immediate survivalist response, which is to hit the gas and get out of the way. So our blood pressure may rise and our heart may beat a little faster, but we don't process our impression consciously.

This sort of unconscious registration is happening all the time—in response to everything we encounter in our immediate environment. It's part of the way we adapt. We always "know" more about ourselves and others unconsciously than we could possibly know consciously. This is true in every part of our lives, whether we're interacting with an analyst or interacting with a friend, spouse, or colleague.

Unconscious Communication

The crucial issue, of course, is what happens to this constant stream of unconscious perceptions? Do they have an effect on us? The answer is deceptively simple. If they're important to us, *we talk about them.* They emerge in the course of our dialogues, our dreams, our fantasies, our self-addressed monologues. This is why a shift from studying the human mind in isolation to investigating its outputs in language can give us a first approach to a science of the mind and its adaptive efforts.

One might ask, of course, how we can talk about things that

have never been conscious. This answer is not quite as simple. To be unconscious of something is quite genuinely not to know about it. Thus, we can't talk about unconscious perceptions directly. We talk about them by means of *analogy*. That is, we allude to a situation already conscious or easily made conscious that is emotionally similar to our unconscious impression and our attempts to adapt to it.

This capacity for unconscious analogy is part of our psychobiological heritage. In a sense, it just "happens to us." We exploit the ambiguity of language without even realizing we're doing so. I generally call these analogies *encoded images*. They tend to appear in the form of stories. Virtually all narratives have two meanings—one direct, as stated, and the other unrecognized, encoded.

To be sure, the idea that people use symbols to stand for unconscious experience is one of the cornerstones of both Freudian and Jungian theories. But the pioneers of psychoanalytic science were attempting to delineate universal images, whose appearance in a client's productions signal the presence of a typical infantile conflict or an archetypal cultural one, which require individuation to be integrated. I don't dispute these findings. I'm saying something different. I'm saying that *immediate* unconscious experience, which is highly personal and individual, is invariably represented by an allusion to something already conscious.

For example, let's go back to the honking horn incident. We felt angry about having been forced to attention, but we didn't consciously process that impression. What happened to it? How do we adapt to an emotionally charged situation that we never registered consciously?

If such incidents have strong emotional implications for us, we adapt to them unconsciously. Later that night, we may accuse a good friend of being pushy and making us feel stupid. The context may even warrant the accusation, but our unconscious impression has dictated its form and intensity. It's served as a vehicle for implications we didn't recognize consciously. We're still trying to adapt to them without realizing we're doing so.

Someone else might represent the experience differently, perhaps telling a friend over lunch about the plot of a TV show in which someone was caught sleeping on the job. Encoded images are *not* universal. They can't be looked up in a symbol dictionary. They are highly individual, embedded in a person's everyday environment, and commandeered as a vessel for adapting to and communicating the unconscious conflict being experienced.

We use encoded images in all forms of emotional interaction, but their nature is particularly clear in the intimate dialogues of a psychotherapeutic relationship. Patients' unconscious experiences of their therapists' interventions are consistently represented or encoded in the stories they tell about other things.

I'm using the word *story* deliberately. Secondary meaning is apparent only in narrative communication—when we describe events, incidents, or happenings real or imagined. When we are explaining things, making rational arguments, philosophizing and intellectualizing, we are not using imagery, and our words express little more than conscious meaning. (This is perhaps one reason that people who restrict their conversation to explanation and exposition strike us as emotionally distant.)

Illustrating the Process

Two rather simple illustrations will help to clarify this further. A therapist had inadvertently extended a client's session by ten minutes, but the client waved him aside. He said it didn't matter one way or the other. Nonetheless, the client registered a number of perceptions unconsciously that were in conflict with his conscious perception. He began the following session by telling his therapist about a boss who had kept him overtime, but hadn't paid him for the extra work. He said he'd felt trapped and exploited, and added that he was thinking of leaving his job.

Although the client wasn't consciously aware of these implications, he was adapting unconsciously to the disturbance in his therapeutic environment—in the ground rules that defined his therapy. He was contending with the unconscious meanings of his therapist's lapse by talking about an analogous experience that was fully conscious. Because these implications had never been consciously registered, they were inaccessible to him directly. And they would remain unconscious unless he had occasion to connect the image of "overtime" with the conscious "trigger" that had evoked it.

Recall that the client had waved the therapist aside. The extension of time didn't matter, he said; it wasn't important. No harm done. This is an example of an intellectualization, and it is essentially devoid of unconscious meaning. It is the client's *narrative* that is two-tiered. On the conscious level, of course, it is simply a story about a problem at work. These surface issues and meanings are directly stated and quite valid. But the same words and language convey a second message, dealing with an issue very different from the consciously identified problem—the attempt to adapt to an unstable therapeutic environment.

Let's take a look at encoded communication in everyday life—a simple example of double meaning in a social exchange:

Ann had been dating Harry exclusively, but without an expressed commitment from him, for the better part of a year. In an attempt to push Harry toward some kind of declaration, Ann asked him whether he would feel upset by her dating someone else from time to time.

Harry reacted in a characteristically rational manner. He said if Ann felt interested in someone else, she ought to satisfy her curiosity. "For example," he said, "when I was away on a business trip last month, I wouldn't have minded if you'd gone out to dinner or a movie with someone else, if that's what you wanted to do. I think our relationship is strong enough to survive a dinner out with other people once in a while."

Ann felt irritated by Harry's answer. It was objective enough to indicate indifference, and it made her wonder whether he, in fact, was seeing other women. She dropped the subject and soon Harry began to talk about his new job situation. "One of the things that drove me nuts about my last job," he began, "was the fact that I just didn't feel like I was taken seriously. I mean, I'd go away on a trip, and come back to find someone else in my office, or some piece of my job parceled out to someone else. I really hated that."

Harry's last statement is a good example of unconscious communication. Even though he had changed the subject of their conversation on the surface, Harry was still reacting to Ann's original question about seeing other men. Only now his reaction was twofold. He had already responded to Ann consciously, but now he was unwittingly conveying his unconscious reaction by analogy.

Harry's conscious image of himself as both tolerant and sure of his relationship with Ann was at variance with his unconscious perception of the implications of Ann's question, which cast doubts on her commitment to him. His fear of not being appreciated and of being replaced rendered this impression unconscious and led to a reaction that was encoded in his double-meaning story about his job. Harry was strongly suggesting, without consciously realizing it, that Ann not take the advice he'd just given her.

If this story sounds somewhat inconsequential, it is also instructive. The attempt to protect one's sense of emotional equilibrium occurs at many levels. It need not lead to the kind of behavior that might be defined as pathological. Ann's question was a form of emotional blackmail common enough in everyday social interaction. Another man might have recognized her motivations consciously and responded to them directly. But for Harry, the threat of displacement and humiliation was highly charged, and his perceptions of that possibility were processed outside his awareness.

Consequently, he seemed unmoved by Ann's unstated agenda. She'd expected him to recognize both her implied threat and her veiled request for commitment, but Harry responded only to her surface question. And he did so in a single-meaning, purely intellectual way. However, that rational response was immediately followed by Harry's unconscious reaction to the emotionally charged context. The shift to unconscious registration and processing enabled Harry to adapt to his immediate context without having to deal with his sense of conflict.

In fact, the job that Harry was remembering had been an interim position. He had taken it as a stopgap while he continued to look for "the real thing." Indirectly, Harry was telling Ann that he not only wouldn't tolerate displacement in the relationship, but had no intention of making the commitment she sought.

The following week Harry accepted a blind date arranged by a friend who knew that he was seeing Ann. This, in general, was Harry's pattern. The price Harry paid for remaining unaware of his fears was an inability to settle down, either vocationally or romantically. He handled both his fear of commitment and fear of being replaced by increasing his options at the expense of established situations.

Notice that Harry ultimately responded to Ann's request/threat with action. Although his initial attempt to adapt occurred at the level of interactive communication, the emotional implications of his conflict created enough tension that he sought behavioral relief. He acted out the unstated agenda of Ann's question. He turned the tables on her and put *her* in the position of feeling displaced. He let her know that he was "still looking." This is a paradigm for the way in which unconscious communication can, if not interpreted and consciously appreciated, lead to unconsciously driven behavior—an everyday occurrence to be sure.

THE IMPORTANCE OF BOUNDARY CONDITIONS

It seems clear that the division of labor between conscious and unconscious perception helps to maintain psychic equilibrium and stability. A surprising amount of information and meaning is not represented to consciousness simply because the situations perceived will not allow for an unconflicted response. The problem is that once a situation's implications have been processed unconsciously, they almost always remain unconscious and exert their effects without our knowledge.

Moreover, we are not readily motivated to recognize these unconscious implications. They are usually incongruent with our self-image or our usual way of doing things, and, often as not, the conflicts they precipitate are inherent in being human and cannot be fully resolved. There are countless unconscious impressions that are never represented in our narrative communications. They are inconsequential to our emotional well-being. But the simultaneous hunger for freedom and security, apparent in Harry's encoded images, is quintessentially human. To become aware of this kind of emotional ambiguity is more likely to produce anxiety than relief—and relief itself may be a matter of accepting what cannot be changed.

Many years of analytic work have taught me that the yearning for both containment and freedom is fundamental to the human enterprise—at every level of relationship. As such, its implications cluster around many specific therapeutic behaviors that involve boundary issues. Because these implications always involve the particulars of the therapeutic environment—issues of space, time, power, responsibility, control, loyalty—they are fairly consistent from client to client. However, each client attempts to adapt to therapeutic interventions in terms of his or her own life experiences.

By way of the stories they tell, one recognizes, rather poignantly, how the demand for material constants—a therapeutic "frame," if you will—vies with the fear of being trapped. Such parallel responses call into question Freud's understanding of the therapist as a reasonably objective observer in the therapeutic process. Freud enjoyed the luxury of this belief because of his focus on "intruders" forcing themselves through the door between the unconscious mind and conscious awareness.

But, as I've been attempting to make clear, the mind's door opens two ways. The therapist is part of the environment to which a client is adapting. And, as we've seen, this adaptation occurs both consciously and unconsciously. We know this by taking note of the mind's output in language.

Client and therapist are inextricably woven together in a communicative system, and their verbal exchanges reveal both the conscious and unconscious nature of their experiences of each other. Indeed, woven into the patterns of all human communication are our ongoing attempts to adapt emotionally to one another. People who talk to each other—in any life situation—inevitably create an interactive communicative system. Language, in this respect, is revealed as a vital and basic biological function, serving as the vehicle of emotional adaptation at multiple levels of experience.

Thus, a study of communicative exchanges at the level of language can provide observable evidence of natural consistencies and regularities in this process. As a natural system, a verbal exchange is liable to the kinds of questions being asked in physics and the new sciences.

SHIFTING THE PARADIGM

As I've indicated, only certain ways of using language—narrative communication or storytelling—enable the simultaneous communication of conscious and unconscious meaning. This kind of two-tiered communication is an inborn human capacity, derived from linguistic ambiguity, and, as such, should be subject to natural law. That is, its particulars should occur in patterns like the ones that characterize other natural processes.

Because storytelling and intellectualization are external things, they don't fall into the realm of purely "subjective content." They can be observed, measured on a scale of more or less, and analyzed apart from the immediate experience of therapy. Dr. Badalamenti and I took a closer look at these aspects of communication: first, in the controlled setting of the therapeutic system, and later, in the emotional exchanges that are a part of our daily lives.

The initial data on which this book is based comes out of a pilot project that involved ten audiovisually recorded psychiatric consultation sessions. In two cases, the same client was interviewed by three different therapists, so we had an opportunity to see how interaction with different people would affect a client's inclination to tell stories, the way those stories were told, and the potential regularities that lay beneath these surface phenomena.

In the pilot project, each session was transcribed, then scored line by line using sixty hard variables (e.g., Who is speaking? For how long? Self-reference? Change of subject? Positive or negative in tone? Story or intellectualization?). This data was compiled and fed into a graphics program that represented the therapist with one color and the client with another. This gave us an immediate picture of how a session was structured simply by looking at the variations on a simple bar graph.

Once we could see a fifty-minute exchange in a glance, structural differences became apparent. In some cases, a client told the same stories to all three therapists; in other cases, the stories were sparse and many potential narratives went untold. It was evident that the way each therapist structured and conducted his or her particular session changed the way these stories emerged. This finding may seem intuitively obvious, but it was worth documenting on paper. It also opens a larger window of observation, as this book will indicate.

Our primary interest, however, was not in the differences between the sessions, but in the features they shared in common. How does communication normally unfold between two people in a specific context? It struck us that if we could express mathematically what happens in a therapeutic exchange, we might discover patterns not unlike those that underlie life-changing structures of all kinds.

To develop our more definitive studies, we selected the most workable dimensions among the initial sixty measures we had first defined. Because we were primarily interested in narrative imagery, all but one of our selections involved features that distinguish stories from intellectualizations (extent of narrative, newness of theme, degree of positive imagery, degree of negative imagery, and continuity of thought and dialogue). Essentially, we intended to quantify the shifts of both clients and therapists in and out of the narrative mode of expression—*how* we communicate, rather than *what* the communication is about.

With our five measurable dimensions in hand, Dr. Badalamenti and I attempted a formal mathematical approach to the data of the ten consultation sessions. If we were successful in locating meaningful patterns, we hoped to extend our studies to ongoing, recorded psychotherapy sessions, as well as to emotional dialogues and even monologues drawn from daily life. Each step of the process would determine the decision as to where to take the next step.

The rest of this book is devoted to showing how a scientific approach to narrative and nonnarrative means of communication can afford us a deeper understanding of the way emotional adaptation takes place.

CHAPTER THREE

Why Mathematics?

Reality must take precedence over public relations,
for nature cannot be fooled.

—RICHARD P. FEYNMAN

"SCIENCE," SAID LOUIS ORR, M.D., President of the American Medical Association, at a commencement address in June of 1960, "will never be able to reduce the value of a sunset to arithmetic. Nor can it reduce friendship or statesmanship to a formula." True enough. Why turn pure experience into mathematical abstractions? Why try to quantify any aspect of our emotional lives?

I said in the first chapter that we are accustomed to seeing the therapeutic enterprise through the eyes of Freud, who understood it as a method for eliciting subjective thoughts, fantasies, hopes, and dreams from a client. This idea still gets in our way when we try to unify science and psychoanalysis. We naturally wonder how someone would measure love or anger or infatuation. Science seems too cold, too clinical, too bent on isolating a feeling from its context to do the subject any justice. In our attempt to analyze rather than experience, we may lose all the wonder and spontaneity and mystery of life.

The truth of the matter is that mathematics is useful precisely because it *can't* be used to measure the specific content of anything. Mathematics is not like arithmetic. To paraphrase Madeleine L'Engle, author of *A Wrinkle in Time*: Arithmetic tells us that if you have three apples and multiply them by zero, you get zero. What happened to the three apples you already had?

Mathematics is a very different animal. It describes the form of a thing—its boundaries and how they change over time. James Burke, on the television series *The Day the Universe Changed*, offers the delightful example of a bicycler pedaling his way across France. If we picture the cycler's progress in terms of a graph, the horizontal axis will reflect his accumulated distance. The vertical axis reflects the ups and downs of the road. One can see how the numbers ultimately begin to describe a form—the external trajectory of the bicycler's journey.

Obviously, the form of the cycler's trajectory tells us nothing about the trip he took or why he was taking it. But it doesn't follow that turning his linear progression into a series of coordinates unmakes the joy and spontaneity of his adventure. The fact is that our ability to describe a trajectory across space and time has helped to create science as we know it. Scientists ultimately used such measurements to formulate the law that predicts the rate at which gravity affects any moving object.

Mathematics is essentially transpersonal—to the extent that it was once understood as a form of religious expression. Some cultures constructed their temples by way of sacred geometry, believing that the experience of that structure would align the psyche of a worshipper with a reality outside of itself.

Paul Rapp and his colleagues at the Medical College of Pennsylvania in Philadelphia have been able to demonstrate the very real basis of that belief. The behavior of mathematical systems can be correlated so closely with the behaviors of the central nervous system that dynamical systems theory actually provides a metaphorical language for describing brain electrical activity. This is the kind of work we were attempting to accomplish in the realm of interactive communication.

As psychoanalysts, we are like cooks who think we understand what soufflés are because we've observed eggs and air. Mathematics moves us beyond the naive idea that our observation of cases, our inferences about others' emotions, and our statistical correlations of certain symptoms with certain antecedents are scientifically reliable. Thoughts are as substantial as electrons; they are a form of energy that can change our bodily chemistry, precipitate behavior, alter the face of the earth.

It was Freud's one–time protégé, Carl G. Jung, born a generation after him and witness to the quantum revolution, who first anticipated the potential interface between psychology and hard science:

> *Sooner or later nuclear physics and the psychology of the unconscious will draw closer together as both of them, independently of one another and from opposite directions, push forward into transcendental territory, the one with the concept of the atom, the other with that of the archetype.*
>
> *The analogy with physics is not a digression since the symbolic schema represents the descent into matter and requires the identity of the outside with the inside. Psyche cannot be totally different from matter, for how otherwise could it move matter? And matter cannot be alien to psyche, for how else could matter produce psyche? Psyche and matter exist in one and the same world, and each partakes of the other, otherwise any reciprocal action would be impossible.*
>
> *If research could only advance far enough, therefore, we should arrive at an ultimate agreement between physical and psychological concepts. Our present attempts may be bold, but I believe they are on*

the right lines. Mathematics, for instance, has more than once proved that its purely logical constructions which transcend all experience subsequently coincided with the behavior of things. This, like the events I call synchronistic, points to profound harmony between all forms of existence.

The idea that psyche and matter are different aspects of the same thing is no longer the peculiar concept it seemed when Jung wrote this passage in 1950. The fact that matter can act as both a particle and a wave is central to the branch of science called *quantum mechanics.* This is very much how Jung described the archetypes. He understood the archetypal level of reality as a structural force found in nature that organizes both material and psychological events.

The unconscious need for defined structure and boundaries that appears to be resident in analytic clients can be described as archetypal in this respect. It would seem to direct human perceptions into particular conceptual forms, independent of consciousness and the will. Such reactions are the psychic counterparts of goose bumps and other automatic physical adjustments to a changing environment.

But how does one get at structural laws? Who is to say they exist at all? This is the point of using mathematics. The only way to discover a law is to rigorously observe the predictable—the mundane, ever-present properties of objects or events—and to measure them from moment to moment as they change or remain stable across real time. These measurements become the basis for mathematical models, which can show whether the events we've documented occur in a consistent pattern.

The idea of doing this is as old as Greek philosophy. If it can't be measured, said Aristotle, it's probably a figment of your imagination. (He didn't say this exactly, but the paraphrase is close enough.) Yet, a thousand kingdoms rose and fell without anyone formulating a law of gravity or motion. And

Aristotle's placement of the earth at the center of an unchanging universe has turned out to be a figment of his own imagination. Until Galileo restricted his measurements to the most basic of variables—mass, position, and time—science as we know it scarcely existed.

A story tells of a delegation of sailors who went to the tribunal of the Inquisition in the seventeenth century, when the Church had forbidden the use of modern astronomy as an affront to Aristotelian theory, which supported the Bible's account of creation. The sailors sheepishly confessed that astronomical theory had both simplified their journeys and made their maps more accurate. They hoped that the inquisitors would exempt mariners from the Church's proscription against it. The tribunal considered the problem, spoke with the bishops, sent an emissary to the Pope. Finally, they conceded. They said, "Okay, if the theory works, use it. But don't *believe* it."

This story is no doubt apocryphal, but the point it makes is important. Formal science is not about belief or lack of belief, or about reducing full-blooded experience to numbers. It's about what works and explains best. Scientific law is an abstraction that predicts the effects it describes. Once we can predict specific effects with certainty, we've unlocked a secret that can change the way we live our lives. Experiences that could not otherwise have been imagined become ordinary events—like turning on a radio, talking to someone who lives across the world, or driving through a tunnel underneath a river.

After all, does knowing how the human eye works diminish the distinct pleasures of gazing at a sky full of stars? Ask anyone who has seen the Pleiades through a telescope or recovered sight after cataract surgery. Knowing how our feelings work is no different than knowing the biological mechanisms that make vision possible. What if that knowledge could produce the equivalent of emotional contact lenses, enabling us to participate in life more fully, more insightfully?

Of course, the actual process of science is always somewhat tedious. Any attempt to discern the underlying structure of predictable events requires that we begin with the self-evident—variables so obvious and ever-present that we take them for granted. As indicated in the previous chapter, we began our investigation into the structure of communicative interactions by transcribing data from ten psychiatric consultation sessions. In our first probe, we measured the most apparent variable we had: Who is speaking, and for how long?

Indeed, this particular aspect of our data is self-evident enough to seem thoroughly unrewarding. The exact amount of time a person spends talking may not seem to be of much consequence. We talk about people being good listeners or the irritation of being interrupted in midsentence, but such informal impressions are generally useful only in making decisions about dinner invitations, marriage proposals, and the like.

This is exactly why the subject is important. We know *nothing* about this simple dimension of human communication save for immediate subjective experience. Yet the alternation of roles and the length of time someone speaks serve as the temporal backbone of all possible conversations. It therefore makes sense to ask: Is there some consistency to this dimension? Do our conversations have a stable base that transcends the individual dialogue, or are the variables completely random? And if there is some pattern to speaker alternation, what effects does it have on how we relate and interact with others—within and outside of therapy?

These are the kinds of questions mathematics can help us answer. Science not only reveals the laws and regularities of nature, it also guides how we, with our gift for adaptive knowledge, can manage nature.

CHAPTER FOUR

Initial Data

Whether man is disposed to yield to nature or to oppose her, he cannot do without a correct understanding of her language.

—Jean Rostand

Before we turn to the mathematical aspects of the research, let's get a feeling for the data by taking a quick look at three of the consultations (disguised) we used for our pilot investigations. All ten consultations were recorded for presentation at a professional conference. The clients had agreed to participate because their personal therapies were not going well. They had hopes of either resolving the impasse or eliciting a second professional opinion.

The three excerpts feature the same client, a woman I'll simply call Client 1, who was in her late fifties and suffering from depression. She saw three different therapists: Therapist A, whose method of interaction was conversational and empathic; Therapist B, whose method was more interpretive;

and Therapist C, whose interventions were probing and aimed at clarifying the client's statements. The three interviews took place several months apart in the order given here.

THE FIRST INTERACTION

Therapist A: How was it when you were little? Did you feel unprotected and needing care? How was it with your family, as a little girl?

Client 1: (*speaking before Therapist A has quite finished*) Well, my father was an alcoholic, and it was during the Depression, and, ah, I was very, very allergic, and going into shock a lot, and they didn't know too much about allergies at that point in time, and they kept giving me things repeatedly that, or having things around in the environment that made it worse, and . . .

Therapist A: So there was a lot of bad care even there.

Client 1: Right. So I . . .

Therapist A: And, if your father was an alcoholic, he wouldn't make you feel very secure, either.

Client 1: No. And, ah, then my mother left, we left when I was about eight.

Therapist A: Left the home.

Client 1: Right.

Therapist A: Were you the only child?

Client 1: I am the only child, and she's an only child . . .

Therapist A: I see . . .

Client 1: . . . and she has been ill all of her life.

Therapist A: Oh, has she? What, ah?

Client 1: Well, she had high blood pressure, and a heart condition—still does, but umm . . .

Therapist A: So there's your caretaking self (*therapist laughs, so does client*), forever needing to look after your mother, too.

Client 1: Right.

Therapist A: Maybe she wasn't very able, with all this illness, to give you a feeling of being cared for and looked after. Do you feel that?

Client 1: No, I don't think that I did. I only felt—I felt loved with strings attached . . .

Therapist A: Unh-hmm. What were the main strings?

Client 1: . . . and, ah (*speaking over therapist's voice*), if I performed well . . .

Therapist A: (*speaking as client speaks*) Oh, yes, you would get loved . . .

Client 1: I was dancing as a little child on the stage, and so forth, and when I would feel I did a good performance, then I felt, I mean, that's what I felt was the only thing in life that made me feel cared about.

Therapist A: She appreciated your dancing? She was proud of your dancing?

Client 1: I think that she probably had wanted to be a dancer, and I was sort of pushed into it at three and a half because I was pigeon-toed and knock-kneed, and just went on.

Therapist A: So you were to be her dancing self in a way?

Client 1: I guess so.

Therapist A: Yes. So you tried very hard to dance well, to do good performances.

Client 1: No. I tried . . . I was rather bad so I wouldn't get the parts. I remember quite often doing something wrong . . .

Therapist A: (*before client has finished*) Oh, you didn't want to do that, you felt pushed.

Client 1: Right. And I felt at the end of the . . . she had a heart attack when I was fourteen, and at that point I was having a private lesson every day, and I would be crying most every day.

Therapist A: Oh, I see. So it was not a happy choice of . . .

Client 1: Preten . . . ah, well, it was a very demanding ballet teacher.

THERAPIST A: It certainly is, one has to really love it . . .

CLIENT 1: (*before therapist has finished*) . . . every day, work every day . . .

THERAPIST A: . . . and want it passionately to put up with the strains of, ah, ballet training.

CLIENT 1: Right, right. So my life wasn't normal because of allergies, and because of being pushed into being a Broadway child.

THERAPIST A: Hmmm-hmmm, yes, which you didn't want. And feeling that your mother wanted this, and that she was so ill as well.

CLIENT 1: Umm-hmm.

THERAPIST A: So that, in part, not only would you not feel very secure, but you felt you had to keep caring for her, and pleasing her. That must have felt like a big task.

CLIENT 1: Right, and then I think I loved my father more, I hate . . . it seems like I felt hatred toward her up until, say, the last few years.

THE SECOND INTERACTION

THERAPIST B: Let's see what happens. Let's just see where your thoughts go . . .

CLIENT 1: (*sigh*)

THERAPIST B: Whatever occurs to you . . . I'm going to let you sort of . . .

CLIENT 1: I'm in therapy, you know, with Dr. Hardy [*her private therapist*]. I'm seeing him because I've been depressed for the past two years, at least that long, maybe longer. (*pause*) It began, I guess it began when I fell ill one winter with the flu, then my kidneys got infected—not that the doctors realized what it was right off. I was pretty sick. They didn't know what I had. They put me through all those X-rays and tests. It was pretty awful. I think the X-rays did me some harm. I don't know. (*pause*) It was like it was when I was a child, when I had so many illnesses and was going into shock all the time. I felt different from everyone else. The doctors did a lot of that other thing, too—you know, what's it called?—fluoroscopy, looked at my stomach and my bowel. They weren't sure where the problem was. I thought it was with all those procedures. (*pause*) I had a friend who had these bad headaches and she got real depressed with all of the tests they did on her, brain-imaging and the rest. She hated being in that machine. She ended up attempting suicide with some pills. I hated being examined, too. It was at the university hospital and people would come and go, stay there and watch the doctor examine me. It was terrible. It's no way to treat a patient.

THERAPIST B: Um.

CLIENT 1: When I was very small, about three and a half, I was given dancing lessons to cure my pigeon toes and knock knees, and so I was dancing about every day until I was twelve. So that made me different from everyone else, too. I didn't have a normal childhood, playing. I was every day in dancing school. Along with that, I felt that my mother's love was never given freely. It was only if I performed well. Onstage. If I performed well. And it seemed like she only gave me love when . . . if . . . and there were strings attached to this love, you know. It may not be true, but that's my perception of it.

THERAPIST B: Yes, yes.

CLIENT 1: So I felt that, continuing through my life as well. But . . . I'm only, I'm only a viable . . . I'm only, I have an identity when you can see what I've done or I know that it's performed well. Like, grades on a test or something of that nature . . .

THERAPIST B: What about like right now, where you're performing . . .

CLIENT 1: I'm not performing as well as I could be or should be at my new position.

Therapist B: I mean, at this moment, here, in this interview . . . where you're talking about . . .

Client 1: Oh, here . . . Oh . . .

Therapist B: There's a parallel . . .

Client 1: Oh, no, no . . . I feel perfectly all right now. You're not judging me in any way, shape, or form. I don't feel that I'm being intimidated or . . . like someone that might be close to me . . . ah . . . someone that I might want their affection or love or approval from. Not that I . . . don't want your approval, but I mean . . . it's not, you know, it's a different thing, I think.

Therapist B: As I've been listening to you, although you've said that performing wasn't really what you wanted for yourself, I don't think performing is entirely negative for you. I think you're also saying that you like to perform, and that has something to do with your agreeing to this interview.

Client 1: Oh, yeah . . . you're right.

Therapist B: You've also been talking about doctors who don't know what they're doing and about being scrutinized, examined, and exposed to a lot of people, feeling that all of it was harmful to you—that something like it caused your friend to try to hurt herself. It seems to me that while, on the surface, this has something to do with your depression, you are at the same time trying to tell me something else—something that refers to what's happening right now, where you're also being scrutinized, examined, and once more in front of many people. From your stories, it's clear you feel you're being harmed by this procedure, like you felt you were harmed when you were examined before, and that you don't like it at all. You're telling me that what I'm doing is no way to treat a patient.

Client 1: (*pause*) You know, I've taken up horseback riding . . . the last four weeks. Again, something that I enjoyed before and I had stopped for about nine years. The physical thing that . . . I mean, my body is not really all that equipped to do things that I used to do, and some of the techniques have changed, but it does give me a . . . even in the three weeks I've been doing it again, I've gotten some muscles back and I do feel a sense of . . . even though I still do everything not that, you know, great, but it does give me a sense of the performance well done . . . Even if, you know . . . other people in the class, like little kids in this class, or whatever. And the teacher's always correcting, you know, but I still feel a sense of accomplishment on my own part, because I can see the difference from week to week, the muscle strength, or whatever, so . . . Guess I'm giving myself approval by continuing along these lines of doing something like this.

THE THIRD INTERACTION

Client 1: We always go back to the childhood. I think people try to blame their present problems on their childhood.

Therapist C: Are you trying to do that?

Client 1: Sometimes.

Therapist C: Tell me.

Client 1: I've tried to forgive my mother and father for not being perfect, in that my not having a normal childhood, whatever that is, but I was very ill as a child. I had asthma, and I kept going into shock and they didn't even know what it was, and I'd get blue and stiff and whatever, and it was just in the days when they were learning about allergies and they put me in a steamer tent in hot and cold water and . . .

Therapist C: This is something you still recall vividly.

Client 1: I still have a phobia about heat coming at me and certain things, and I've had somewhat of an aversion to my mother. I don't know exactly why. My father was an alcoholic, and I don't recall too many weird episodes like some children, but I know that there, I do recall some . . .

Therapist C: Mm-hum.

Client 1: . . . and she left him when I was eight, but I was very knock-kneed and pigeon-toed and I was put into dancing school, and my whole life was in dancing school. I was not allowed to play and every day I was in the school, and, ah, I always felt that there were strings attached to my mother's love. If I performed well, then, that was okay . . .

Therapist C: Mm-hum.

Client 1: So . . .

Therapist C: You felt that you had to do it for *her*?

Client 1: I don't know what for, but I felt that the only way that I was accepted was if I performed well.

Therapist C: Um-hmm.

Client 1: I don't think I, I was intelligent, but I kept getting into trouble in the early grades, because I could read before I went to kindergarten and I think I tried to get attention even in the early years and was happiest when I was the center of attention . . .

Therapist C: Uh-um.

Client 1: . . . whether, I don't really know what it stemmed from, but in a way I suppose that's carried over, too.

Therapist C: The way you are describing it, it sounds as if you *shouldn't* have or what?

Client 1: Shouldn't have been the center of attention?

Therapist C: Uh-um, or shouldn't have wanted that, or . . .

Client 1: I don't know. I don't think it's normal, my childhood, you know, not having . . . not, I didn't really have a normal, I had some friends, but it wasn't, well, being very ill, and then being all the time in dancing school and being different and set apart and I would be taken to the studios and I would do things wrong so I would not be chosen. I remember very vividly doing (*laughs*), just doing something wrong so I wouldn't be the star they wanted me to be.

Therapist C: Well, what would that mean?

Client 1: So that's a dichotomy about me wanting to be the center of attention, but I was bad so I wouldn't be.

Therapist C: Mm-hum.

Client 1: I don't know.

Therapist C: You mean you both wanted it and also felt uncomfortable with getting it?

Client 1: I think so, in retrospect. I think that it does seem strange, and even in my later life now sometimes I'll back off from success . . . not really do something wrong, but . . .

Therapist C: Not take full advantage of what's available by way of success.

Client 1: Right, right. I'll stop short, in other words, for some strange reason.

Therapist C: Does that create some problems that you can't really allow yourself to, you know, sort of to reach the maximum?

Client 1: Yes, I feel very . . . on the one hand, I feel better than or less than. I really don't ever feel equal.

Therapist C: Ah-um, or at the right level?

Client 1: Right, right. Much of the time. I became a heavy drinker when I was in my late twenties and I really had kept myself away from alcohol because of my father, even before I knew it was maybe hereditary. (*laughs*)

Therapist C: Mm-hum.

Client 1: I started drinking at twenty-eight and I've stopped now. It's more than ten years, but, ah, I think when I was drinking that I did feel that I was more equal to rather than better than or less than.

Therapist C: Mm-hum, tell me about that a little bit more.

Client 1: What part of it?

Therapist C: More equal.

Client 1: (*laughing lightly*) Ah, I think from my going to AA, I think that a lot of alcoholics feel that way. They feel very sensitive and very, ah, less than or better, you know . . . it seems to be a common denominator I've found throughout the years.

Therapist C: You mean, some areas better than the rest of mankind and some areas worse?

Client 1: Well, the chemicals, alcohol included, make . . . made . . . makes the alcoholic feel everything is all right in that imperfect world out there and then we can cope. (*laughs*)

Therapist C: Um-hmm.

Client 1: And so (*pause*) I didn't drink very long, but I've been under care for depression for some years. I get very depressed and . . .

Therapist C: That's what you're in therapy for right now?

Client 1: Uhhh—yeah, depression and, I think it goes along with the feeling "less than."

Therapist C: Mm-hum.

Client 1: Maybe, I don't know.

Therapist C: Tell me—less than *what*?

Client 1: Less than I should be or other people or the way it should . . . well, less than I should be—maybe because I was pressured to be the best at an early age.

COMPARING THE SESSIONS

I deliberately chose excerpts in which the client recounted some of the same life experiences—childhood illness, dancing school, an alcoholic father. Sometimes she used the same phrases in each of the three consultations. However, the speaking patterns in each one helped to shape the way such events came up and were understood.

Superficially, of course, we can get an immediate impression of these patterns. Therapist A spoke even more than the client, a style that held true throughout the entire interview.

The client's replies were somewhat clipped, and her narrative images were undeveloped.

The interview with Therapist B was less like a conversation. In fact, Therapist B spoke very little in the first part of the dialogue, then spoke a great deal in a single interpretation. The client held the speaker role for a long time, seemed more articulate, and told more stories.

Therapist C encouraged the client to elaborate on her statements, and he didn't interrupt as often as Therapist A, but the client's replies were somewhat intellectual, and there are few stories to speak of.

Consistent across the three cases are images suggesting an unconscious response to the exposed nature of the consultations. The clearest in this regard are the ones about her ambivalence as a performer, and the one elicited by Therapist B, about confused doctors who X-rayed her and harmed her in the process. When Therapist B interpreted these images, the client produced a positive image about learning new things and being pleased with herself, despite being "corrected" by a teacher.

Intriguing as these observations are, they are simply impressionistic, and they are focused largely on meaning. Unless we know how a dialogue normally proceeds, or whether it has a regular pattern at all, we won't know what the variables are or what they mean. This is why we can't begin with *content* in our attempt to search for patterns. On the surface, there is no evident pattern that can be abstracted with any assurance. Indeed, all ten sessions had their own distinct character, and, despite some consistency in the client's productions, there was no apparent rhythm or order to the process of speaking in any of them.

On the other hand, *all* dialogues have a common, self-evident property—grounds on which they can be measured and compared. They always involve two people speaking in turn.

One can determine exactly how often and for how long either participant spoke, and search for consistencies and signs of a structural pattern in these measurements. By quantifying this aspect of the dialogues, we get an entirely different picture of the situation. Beneath the surface of these exchanges, there is indeed a mathematical regularity, which suggests that human communication shares common ground with other natural physical systems.

As we outline the research we did, we'll be talking about different ways of ascertaining mathematical regularity, and speculating on the hidden structures and laws that evidently underlie even the simplest of human exchanges.

CHAPTER FIVE

It's About Time

Time is the wisest of all counselors.

—PLUTARCH

JUST AS COGNITIVE science studies the nature of mental tasks and the processes that enable them to be performed, we are proposing a science of emotional cognition. We understand this science as a branch of biology, because its domain is human adaptation. Its purpose is to study the communication and processing of emotionally charged stimuli. Psychoanalysis, in these terms, would be defined as a branch of the science of emotional cognition, focusing on the nature of *unconscious* emotional perception, adaptation, and communication.

In Freud's lifetime, psychoanalysis was a kind of fledgling science—a set of theories based on direct observation. Most sciences begin this way. But, as Descartes once suggested, sense impression can be sense deception. At one time, everyone "knew" that heavier bodies fall faster than lighter ones—until falling objects were studied in a vacuum.

Statistical science introduced the concept of measurement into psychoanalytic research, but the summary nature of statistical results affords them only limited utility. I mentioned earlier the correlation between liking one's therapist and a good therapeutic outcome. Although the two phenomena may occur together more often than not, we don't know what causes the correlation.

In the same way, it can be shown statistically that high sugar consumption is linked with hyperactive symptoms. Again, the correlation may be accurate, but whether sugar in some way contributes to hyperactivity or whether hyperactive children are simply less able to delay gratification becomes a matter of pure speculation. Broad connections just can't illuminate the basis of a regular process in nature.

One might say that psychoanalytic research is in a transitional period—evolving from a statistical science to a formal one. Such evolution generally occurs when observation becomes more rigorous, more attuned to process than product. One selects an observable linear phenomenon, reduces its particulars to numbers, then records their progression over time. The resulting moment-to-moment data is called a *time series.* When a time series has enough inner regularity, it can be expressed by an equation that, in effect, predicts the results obtained.

A shift from impressionistic science to mathematically grounded research usually revolutionizes the field in which it occurs. Recall our intrepid bicycler and the graph that describes his movements across France. Replace the bicycler with a planet and you have an idea of what Tycho Brahe did when he measured the nightly movement of the planets in space and time.

From Brahe's moment-to-moment data, Kepler recognized those aspects whose regularity could be captured mathematically. He developed a model that not only predicted the results Brahe had obtained, but would predict all future results as well. We now accept his model as the laws of planetary motion and elliptical orbit.

Another example of such a model is the law of gravity, which is expressed in an equation that predicts the rate at which all bodies fall to earth. Although we are accustomed to thinking of such laws as applicable only to physical objects, we were hoping to accumulate data that would establish regularity in human communication.

We began, however, with another kind of equation, one that deals not with predictable certainty, but with likelihoods or probabilities. Such mathematical statements describe regular patterns, but they do not establish laws of nature. We figured if we got meaningful results with less formal mathematical models, we would attempt to create more rigorous ones and search for predictable laws of communication.

CRUNCHING THE NUMBERS

For our first study of the ten consultations, we turned to a relatively simple and basic aspect of every communicative exchange—*who speaks in turn and for how long*. This is a foundation for all conversations, so we theorized that it might have some regular structure and predictable effects as a dialogue unfolds.

We had two considerations in mind: first, the pattern of speech durations made by the client and therapist together—the *C/T system*; and second, their *individual inclinations* to speak and be silent. That is, we wanted to study the client and therapist both conjointly and as separate speakers. To do this, we simply recorded the number of seconds that a respective client and therapist spoke in turn. This was the data we then studied.

The *switching of speaker roles,* which is a *systemic or conjoint measure,* has properties in common with a measure like the Dow Jones Index. The fluctuations over time have a degree of stability, in that they adhere to a certain pattern. The way things are at one moment is most likely the way they'll be the moment after. But such phenomena also involve random variation. There are sudden and dramatic shifts in the pattern.

So we used mathematical models originally intended to study processes that have both fixed and uncertain aspects—not only economic indicators, but traffic patterns, the shifting of populations, sun spots, and so forth. Such models are known as *stochastic,* from a Greek word that means "aim" or "guess." The name indicates the nature of the models' design. They characterize the underlying pattern in a process that has an element of chance. As I said earlier, by measuring the likelihood of chance events in a time series, we have an indication of *probability* rather than *predictability*—not a law, but a broad regularity, a pattern that involves both certainty and uncertainty.

The stochastic models we used for our measurements are called the Box-Jenkins models, after their inventors. They permitted us to ascertain both the rate of stability and the rate of instability in a communicative system. Again, *stability* means that a system changes in a fairly predictable way and not too suddenly. In the inimitable words of Yogi Berra, it's déjà vu all over again. Most systems in nature are inclined to stay where they are—to adhere to a given pattern. In terms of our data, stability means that whoever is speaking at a particular second in a session is likely to be speaking the next second as well. Disturbances in a system, called *shocks,* disrupt an established pattern. The larger the measure of shock in the system, the greater its *instability.*

We wanted to know whether client/therapist dialogues had a stable pattern insofar as switching speaker roles was concerned. Would all the dialogues show the same pattern? If so, were there individual variations? Perhaps there was no pattern at all.

In a client/therapist system, of course, the switching of speaker roles involves two distinctly individual propensities.

We might call them *holding the floor* and *interrupting*. The actual switching process, therefore, is unique in each consultation. It involves not only the individual proclivities of client and therapist, but the effects they have on each other. For this reason, we had to determine whether there was some consistency to each session before we could run the entire series. Without this consistency, it would be impossible to establish a pattern for each therapeutic couple and the Box-Jenkins models would not apply.

It turned out that each session did indeed show this kind of consistency. Who spoke and for how long was pretty much the same for any segment of these dialogues. This property is called *stationarity*—for obvious reasons. Once established, a pattern endured. We could therefore compare all ten sessions and use a Box-Jenkins model to make sense of the data. We wanted to create an equation that would describe the degree of stability and instability in a given C/T system—that is, a given system's resistance to change along with its vulnerability to destabilization.

Thus, we expected our stochastic measurements to tell us one of several things—that the switching of speaker roles was:

1. random and unpredictable, without pattern;
2. stable and predictable, despite surface differences;
3. generally unstable, but in some regular way; or
4. stable, but easily affected by shocks.

LOOKING AT THE RESULTS

As it turned out, our measurements confirmed the second possibility. All ten sessions, despite having their own unique characteristics, showed a great deal of stability. This, in itself, suggests that the alternation of speaking roles, which takes place with little conscious thought, is not a disordered, random, or chaotic process. Although the relative stability or instability of the system varied from matchup to matchup, the dialogues were governed by an underlying regularity that could be described mathematically.

This finding—of both consistency and individuality—was very appealing. It gave us a way to draw some new hypotheses about our data. We could infer a deep regularity characterizing all communicative dialogues with enough room for variations that would distinguish one C/T system from another. We could establish an *index of stability*.

Recall that we were interested in two basic factors—a system's resistance to change and the likelihood of disruption. We measured resistance to change by noting whether the system was dependent on prior values—that is, whether the present speaker would continue to speak. The extremes of our findings were illustrated by two of the sessions excerpted in the previous chapter, those with Therapists A and B.

Although this is not always the case, one can see the difference even at a surface level of observation. Both sessions had stable patterns of their own; however, the session with Therapist A was the most vulnerable to disruption and, hence, the least predictable of the series. The session with Therapist B was the most resistant to change. Stability in switching speaker roles clearly resulted from the therapist's allowing the client to speak for long periods of time.

The measure of instability in our model is especially interesting, because it cannot be observed on the surface of the dialogues. The speakers themselves are unaware of shocks to their communicative system. Disturbances in a pattern are revealed only by mathematics. Thus, it is notable that the pattern of speaker duration, whatever its characteristic structure, was usually broken by the kinds of communications I've been defining as encoded images—emotionally charged unconscious commentaries on the therapist or therapy itself.

That is, destabilization rarely occurred with the introduction of overtly charged subjects, such as sexuality, violence, illness, or death. The typical progression of events involved a sequence of narrative images that conveyed, through stories about other people and situations, rather negative unconscious perceptions of the therapist's efforts. At this point, the therapist would intervene unexpectedly, in a manner that departed from the pattern of alternating speakers already established.

One might speculate that the therapist had unconsciously registered the critical nature of the client's images and acted to disrupt them; however, this takes us too far afield from our data. In any case, the periods of disequilibrium were self-limiting. The system quickly returned to its prevailing pattern of speaker durations, and these interludes were not frequent enough to interfere with the system's general stability. The C/T system regulated itself back to equilibrium.

TAKING STOCK

Our findings, though preliminary, were consistent across the thirty–one exchanges we eventually modeled, both inside and outside of therapy. In fact, the predictions generated by our model closely fit the real-life patterns that emerged in the sessions. Such results suggest the existence of a well-ordered configuration or deep structure in the automatic switching of speaker roles as individuals engage in dialogue. Beyond conscious awareness or control, couples appear to settle into a regular and relatively stable configuration of who speaks when and for how long. And once established, they adhere closely to their pattern, veering off largely when a proliferation of encoded images reaches intensity, without being consciously understood.

Of course, we are dealing here with only one dimension of a therapeutic exchange: how often a speaker speaks and for how long. But there was no reason to expect this highly varied aspect of a dialogue to conform to an exact dynamic as it unfolds across time. Even the sessions that were erratic on the surface showed deep stability with this model. The regularity we found suggests at least one quantified measure that can be used to evaluate the ways in which couples engage in emotionally pertinent dialogues.

We are not, of course, suggesting a connection between the degree of stability in a therapy dialogue and the effectiveness with which it can foster inner psychic change. We were trying to develop some basic tools of mathematically based investigation and did not measure or compare the efficacy of the sessions or therapies we studied.

We did, however, find that in each case, where therapy sessions were concerned, it was the therapist rather than the client who established the speaker duration pattern. Instability usually reflected the therapist's tendency to interrupt the client or to disrupt an already established pattern. Finally, pending further confirmation, one might suggest that a stable dialogue—one in which each speaker is permitted to have his or her say—may enhance a couple's ability to influence each other on deeper emotional levels, even if that influence is unconscious for both parties.

C H A P T E R S I X

Natural Rhythm

The notes I handle no better than many pianists.
But the pauses between the notes—ah,
that is where the art resides!

—Artur Schnabel, quoted in the *Chicago Daily News*

If we can gauge our cultural assumptions by film and TV images, the typical therapist has three basic postures: nodding sagely, taking notes, and occasionally leaning forward from an overstuffed chair to inquire, "And how does that make you *feel*?" This stereotype, whatever its technical merits, exists for good reason. The behaviors constitute our picture of someone who is paying attention. The therapist, as conceived by Freud, is primarily a listener. The client is the one who does the talking—outlines a life history, recounts a problem, explores an emotional response. Even Frasier Crane, the hopelessly neurotic psychiatrist in the TV comedy that bears his name, answers the phone on his call-in radio show by saying, "Hello. I'm *listening*."

Yet, in almost every case we studied in our initial project, the therapist spoke as much as—and sometimes more than—the client! We were intrigued by this finding and decided to explore the data further.

We knew something about how each client and therapist teamed up to create a systemic pattern of alternating or switching speaker roles. Now we wanted to know something about their individual behavior—how long each party held the floor before being interrupted, and how long he or she waited before speaking again. In order to answer these questions, we dropped the Box–Jenkins approach and simply counted the number of seconds a person's utterances lasted. Then we noted how many of those speech events went on for the same amount of time. The net result was a new set of numbers—reflecting how many times a client or therapist spoke for one second, two, three, and so on. These results were placed in a graph called a *histogram.*

Like speaker alternation, the length of time a person speaks or allows another to speak may seem a trivial aspect of dialogue—so trivial, in fact, that we scarcely think about it. We may describe someone as reticent or going on "too long," but otherwise the temporal structure of a dialogue is not something we can easily see. Even if we wanted to, we couldn't conduct a conversation and observe its structural particulars at the same time. The only way to get at this aspect of communication is to measure the duration of each utterance after the fact and to analyze the results mathematically.

So we timed the speech events in all ten sessions and plotted a histogram for each speaker. The points on the graphs represented a speaker's utterances by frequency of a given length. By connecting the dots, so to speak, we ended up with a curve—an abstract picture of a client's or therapist's tendency to speak for long or short periods of time.

Overall, the results surprised us. Despite clear surface distinctions in the ten sessions, the graphs were all quite similar.

The minimum number of seconds a client or therapist spoke was 1 second; the maximum, 253. But in all our consultations, short utterances predominated. In fact, the longer a speech event lasted, the less frequently such a length of time occurred, for either therapist or client. Accordingly, the falloff from short to long utterances was quite sharp. In every graph, the resulting curve descended exponentially.

When something decreases exponentially, its fall-off happens in multiples rather than simple numbers. Perhaps the easiest way to understand the concept is to think about a classroom of 100 children. On any particular day, one or two children may arrive late. If we were to graph the resulting decrease in students who arrive on time—from 100 to 98, we'd hardly get a curve at all. There isn't much difference between a classroom of 100 and a classroom of 98.

But, let's say one of those latecomers was late because he was talking to a friend in the hall. That friend was telling him the secret password to a popular computer game. The latecomer confides the password to two children who sit near him in class. These two children tell two others. These four each tell two more. If the pattern keeps up, before long, the entire class will know the password. The resulting decreases in the number of children who don't know the password is not a matter of simple subtraction. Each decrease is a multiple of the last. If we were to plot this situation on a graph, we'd end up with the kind of curves we were getting for our speech event data.

POISSON CURVES

We were excited by this result, because there are certain kinds of natural phenomena whose graphic representation always results in a negative exponential curve. These phenomena are called *Poisson processes,* after the early nineteenth-century mathematician who identified them. In general, a Poisson process is one in which there are two possibilities. We might call them On and Off, or Yes and No. At any given time, one of those possibilities is actualized. For example, in a huge building like the Pentagon, a certain number of lightbulbs go out every day and need to be changed. So at any given time, a lightbulb is either functional or it's burned out.

If we want to establish a pattern of actual frequency, we might gather data over a representative period of time—let's say a year. At the end of that time, we total the number of times 1 day elapsed before a lightbulb failed, the number of times 2 days elapsed before another bulb failed, and so on. When we plot the answers to these questions, we get our *Poisson curve*, which allows us to predict the likely number of bulbs that we'll have to budget for in our next request for a federal grant.

One of the interesting things about a Poisson curve is the way probability increases as the curve descends. Let's say that the greatest likelihood, where lightbulbs are concerned, is that one burns out in a single day. Thus, if no bulbs go out on a given day, the probability becomes quite strong that one will go out the next day. If it doesn't happen, the probability is even greater that one will fail the next day. The longer the waiting period, the greater the probability of the Poisson event. It's as though a tension were building up, requiring eventual discharge. The tension cannot be sustained beyond a certain point.

Perhaps a more telling illustration than component failure is the project undertaken by a colleague for a graduate course called "Chance and Divination." The assignment was to create a work of art incorporating random possibility. Most of the students came in with "found art" of a visual sort. But my colleague decided to make up a song and ended up with a interesting representation of a Poisson process.

He divided a notebook page into seventy-two sections—one for every ten seconds of a two-hour period. Then he stood on a street corner for the two hours, and each time a car came by within a ten-second period whose license plate contained the letter H, he put a mark in the appropriate section. If he saw two cars with an H plate within the same period, he noted the event with a different kind of mark; if he saw three, he used another.

He based his song on these notations. Its percussion was based on the sequence of the marks and pauses—a single beat, a double beat, a triple beat, and the waiting periods in between. To the surprise of the class, although the data had been entirely random, the rhythm of the resulting song was not. It was regular and stable enough to be defined as music. They could clap along with it.

Like our one-second utterances in the speaker data, the majority of the beats were singles. Out of seventy-two chances, there was only one triple. And, beyond a certain waiting period, a beat became inevitable. Once or twice, a pause went on a shade longer than the class had come to expect from the pattern, and the tension induced a few students to clap too soon.

This is a particularly nice illustration of a Poisson process because it shows how a seemingly random series of Off/On choices ultimately results in a pattern—a pattern replicated throughout nature. It also demonstrates, quite strikingly, how the waiting time between Off and On is an important measure of probability. The longer an event does not take place, the greater the likelihood it will occur—until occurrence is all but certain (for Poisson processes, that is).

Don't Interrupt While I'm Busy Interrupting

Recognizing the Poisson curves in our graphs gave us a way to look at the speaker duration data as evidence of a natural phenomenon, whose deep structure transcends the individual will, allying it with a wide variety of physical processes that seem to be random. Once we understood "starting to speak" and "falling silent" as On and Off states, we could use their frequency to calculate a measure of a person's propensities in a session. The measure is called a *Poisson rate constant* and is inferred from the histogram. The lower the number, the stronger the inclination to continue speaking rather than relinquish the speaker role. The higher the number, the greater the tendency to interrupt the other speaker.

In general, these measures bore out some of the impressions we had gotten from the speaker stability data. Clients, overall, had lower values than therapists, indicating an inclination to keep talking for as long as they could—that is, to assume the role of the primary speaker. The higher values obtained by the therapists reflected their general tendency to interrupt the client. Even those therapists who allowed their clients to speak at length were more likely to interrupt than be interrupted. In one case (the matchup of Client 1 and Therapist A excerpted in Chapter Four), the client's values were even higher than the therapist's, suggesting that the therapist had assumed the primary speaker role. The client was forced to interrupt just to get in a word or two.

When we extended our research to other kinds of dialogues, we found that whenever both speakers obtained high On/Off frequencies, they produced a system characterized, like the one with Client 1 and Therapist A, by short bursts of speech. Speakers silenced one another in midsentence, uttered a sentence or two, then were silenced themselves.

By recalling the class members who clapped too soon when a pause in the performance was longer than expected, one might infer a literal tension that accrues to the need to speak in a conversation. The fact that the likelihood of interruption builds as a Poisson process suggests that tension increases as time elapses. Therapists who allow their clients to speak for long periods of time may have trained themselves to hold a natural tendency at bay.

Characterizing Therapeutic Speaking Patterns

Once we had the speakers' Poisson rates, we could define the probability of each speaker interrupting the other after every second of elapsed speaking time. In this way, the speaking parties emerged as distinct individuals, each of whom handled communicative tension in his or her own way.

This became particularly clear as we combined our individual probability results with the results we'd obtained on the two-person speaker stability. Recall that we had defined stability in terms of speaker continuity. For each line of dialogue, we put down the number 1 if the client was speaking and the number 2 if the therapist was speaking. A series of numbers was considered stable when the previous number usually predicted (was the same as) the next—that is, when the same person continued to talk. Stable systems, thus, inevitably correlated with a greater number of long speech events. This correlation proved helpful in the long run. It gave us a way to distinguish between highly active therapists.

For example, another matchup in our series, Client 2 and Therapist D, was much like the one with Client 1 and Therapist A, in that the therapist had assumed the role of primary speaker. However, unlike Therapist A, whose comments were brief and elicited brief comments from the client in return (12121212121121212), Therapist D tended to speak for long periods of time—until the client was able to interrupt him (11122222222212122222). Thus, the system as a whole was more predictable, had more stability, than the one involving Therapist A.

This distinction becomes important if we hypothesize that role reversal in the therapeutic dialogue, as indicated by Off/On frequencies, suggests a therapist's difficulty with communicative tension. One might even theorize that this is a quantified indication of countertransference—the therapist's unconscious and inappropriate involvement with the emotions, experiences, or problems of the person undergoing treatment.

A thoroughly unstable system, like the one produced by Therapist A, tends to preclude narrative imagery from the client, and essentially limits the interaction to obvious, manifest issues. A more predictable system, like the one produced by Therapist D, generates more narrative material from the client, but the therapist tends to avoid its meaning by interrupting. One can see this in the sort of pattern I mentioned earlier—the therapist's abrupt departure from a stable switching pattern whenever narrative imagery becomes intense.

Admittedly, we were working with consultation sessions, which are necessarily different from an ongoing therapeutic relationship. Also, the sessions were designed to illustrate different ways of working with a client, so the therapists had an incentive to play to the camera, so to speak, and may have been more active than they are in their regular practice. Finally, there are factors in a therapist's decision to speak other than the need to discharge tension after being silent for a certain amount of time.

It should be stated, however, that the results we got with the consultations were comparable to those we obtained in the ongoing therapies we studied afterward. They were also similar to the findings for nontherapeutic couples who did dialogues specifically for our project. Listening appears to require a certain ability to hold back an inherent tendency to reply or interrupt.

Nontherapy (Everyday) Dialogues

The nontherapeutic dialogues I just mentioned were collected from three couples who engaged in dialogues about everyday emotional issues. These conversations showed the same deep stability that we had found in the ten psychotherapy sessions. In fact, all three were stable to the point of inertia. Predictability of switching patterns seems to be a fundamental property of emotional dialogues.

We also found that the tendencies of the individuals to keep speaking or to interrupt could be quantified mathematically in a way that distinguished the six speakers from one another. In this respect, the trends that we had discovered for clients and therapists held up quite well for individuals in everyday life.

NAME THAT TUNE

It occurred to us at this point that the probability numbers we'd derived for each client and therapist might reflect an individual characteristic or trait, like fingerprints—*mindprints* or *communication markers*, if you will. We wondered if each person's pattern of speaking, listening, and interrupting—identified mathematically—is a natural part of who they are, as much as their inflection or tone of voice. If this were so, we reasoned that it should be possible to use the communication markers to identify the consultations in which the same individual had participated.

One of the unique elements of our consultation series was the fact that in some cases the same individuals had participated in different Client/Therapist systems. For example, as indicated by the excerpts in Chapter Four, Client 1 was interviewed by Therapists A, B, and C. Client 2 was interviewed by Therapists D, E, and F. Client 3 had seen Therapists A and B. Clients 4 and 5 had each seen only one therapist—Therapists E and F, respectively.

The challenge, then, was to use the speaker duration results to identify the same individual when he or she had participated in different consultations. This pursuit was feasible because Dr. Badalamenti had been working only with the numbers. He had no idea who the clients and therapists were.

Mixed Results

Dr. Badalamenti first attempted to use the speaker data to predict the identity of the therapists. His results were statistically better than chance, in that he correctly identified three out of the four therapists who had seen two different clients. The one source of confusion came about largely because two therapists had similar mathematical profiles. Thus, therapists did have identifiable "communication markers" across their consultations with different clients.

So he turned to the task of identifying the clients who had seen more than one therapist. But this situation turned out to be entirely different. Dr. Badalamenti's predictions were based on his natural assumption that client patterns were preserved in the Client/Therapist dialogue. The reality, however, was that clients did not sustain their communication markers in sessions with different therapists. Thus, his guesses, in all cases, were completely wrong—significantly so.

The discrepancy between therapists and clients was clear even in the graphs. Clients who were interviewed by more than one therapist tended to mirror the curve of the particular therapist they were talking to. Their longer speech events diminished at the same rate as the therapist's, whatever that rate happened to be.

It should be recalled that the graphs summarize only the number of times a given length of speech occurred in a session—*not* the sequence in which one utterance followed another. Thus, the mirroring tendency does not reflect a client's responses to a therapist's immediate speech cues. Rather, it raises the possibility that intangible cues are being exchanged by speaking parties. Perhaps we are sensitive to communicative preference in ways that we don't yet recognize.

More significantly, these results indicate the presence of *therapist dominance* in every case. This conclusion had been foreshadowed by the variability of these clients with respect to the switching of speaker roles in each session, but now it was apparent that the therapist's pattern had also rendered the clients' communication markers invisible.

It should be recognized that the dominance we're seeing here involves the deeper rhythms and temporal structure of the therapeutic dialogue. This is a level of influence that is not consciously controlled or experienced by the participants. Finding the specific sources of this kind of control requires further investigations of therapeutic dialogues.

SOME CONCLUSIONS

We were fascinated by the Poisson structure we'd found in our speaker data. The speaker stability models had been clear-cut. The idea that an active back-and-forth surface dialogue always shows a regular structural pattern at a deeper level is a bit mysterious. However, one would expect a time series based on speaker alternation to be predictable and stable as long as two speakers allow each other long periods of uninterrupted speech.

It's much less clear how two individuals jointly create a consistent pattern of speech lengths—whose graphic representation always takes the form of matching exponential curves. The most that we can say, at this point, is that our conversational exchanges appear to be grounded in structural regularity. Oddly enough, this very regularity implies that the longer a part of the structure is maintained, the stronger the need to momentarily modify it. Stability calls forth instability. Sameness sponsors change.

Actually, this is to say a great deal. As I mentioned earlier, a client's stories in therapy often illustrate a poignant contest between the need for security and the hunger for novelty and freedom. Our data underscores this contest as fundamental to the natural order.

CHAPTER SEVEN

The Five Dimensions

It's got a good beat. You can dance to it.

—ANONYMOUS RATER ON *AMERICAN BANDSTAND*

ONCE WE HAD a sense of the rhythm underlying our ten therapeutic exchanges, we wanted to know more about the conversational dance it supported. That is, we stopped looking at the structural form of the dialogues, and started to work with the actual flow of communications between therapists and clients.

It should be emphasized again that we weren't trying to measure communicative content. We were looking at *communicative vehicles*—the form of the communication, the kinds of words that a speaker used. As containers, communicative vehicles are a kind of boundary condition, whose nature is more fundamental than the message it contains and conveys.

Moreover, the dimensions we chose were measurable. They could be verified consensually, meaning that different people would score the variables in very similar and often identical ways. We rated them every ten seconds (about two typed lines of the session) in terms of degree or amount. They were:

1. presence of narrative imagery;
2. presence of new subject matter;
3. presence of positive imagery (stories about help, care, happy moments, and so forth);
4. presence of negative imagery (stories about harm, illness, frustrations, and so forth); and
5. extent of continuity from one line to the next.

We chose to measure the last of these variables because we hoped it would tell us something about the influence the speakers had on each other, but, ultimately, it had little impact on our conclusions. It seems that the strongest emotional charges and their changes involved narration-related variables. Thus, the remainder of the book will deal almost exclusively with the first four variables.

The aspects of conversation we focused on are observable in almost any kind of verbal exchange. However, in order to simplify matters for our project, we concentrated on the two female clients in our series who had each seen three different therapists. Client 1 was the depressed woman in her late fifties whose sessions with Therapists A, B, and C are excerpted in Chapter Four. Client 2 was a seriously overweight woman of forty-two, who suffered from malaise and difficulties in functioning. Her therapists will be designated D, E, and F.

Of the six therapists, Therapist A was the only female. Each had been trained in a different school of analysis, but all six understood human psychology as a combination of both internal and interactional factors.

THE SECOND GROUP OF CASES

Before we go on, here are excerpts from the interviews in which Client 2 participated. This particular client had recently discovered a twelve-step program, and her agenda in these consultations was frankly evangelical. All three therapists interpreted her proselytizing as an intellectual defense, and I've chosen excerpts that reflect their different ways of attempting to break through it. Therapist D, for example, was intent on challenging the client, interrupting her train of thought by startling her. Therapists E and F were more conventionally interpretive, attempting to give the client a sense of how her behavior made them feel.

THE FIRST INTERACTION

Client 2: (*interrupting*) Well, I'm saying to you that it has only been a matter of weeks that my life has really, finally, after twenty years of being in different kinds of psychotherapy for years at a time, has . . . has life started to change for the better. As for people wondering why I need psychotherapy, yes, if I meet them at a party, and I happen to mention it, they'll say, "Why?" But most likely if I meet somebody at a party, they're also in psychotherapy—I mean, everybody's in psychotherapy, at least in the world I revolve in.

Therapist D: I would think that any reasonable person, anyone not necessarily endowed with a great deal of suspicion, would by now have started wondering whether you're sleeping with your therapist.

Client 2: (*in a high-pitched voice, stifling laughter*) That sounds so traditional to me. I'm sorry, that's funny . . .

Therapist D: Really?

Client 2: Yeah. That sounds so . . .

Therapist D: But you've—this treatment is so quick, it has been so recent, worlds have opened up to you so suddenly.

I mean, what can a reasonable person wonder except that, huh? You don't see why I would naturally wonder that.

Client 2: No, I don't see why you'd come to that.

Therapist D: All right, well, all right. Let's talk about your viewing audience. You'd like for droves of relatively young people starting therapy to come to your therapist, right? You would like to share your therapist with . . .

Client 2: Hmmm, my therapist . . . I don't know. I—what I was, I really was thinking was warning people against becoming as I . . . I . . .

Therapist D: You want to warn the multitudes away from people like me, right?

Client 2: I'm presuming—I'm presuming that you're someone—that's . . .

Therapist D: (*big laugh*) Hah-hah-hah! You've got to attach significance to a statement like that, haven't you?

Client 2: Well, I hope I didn't just make an insignificant statement . . .

Therapist D: (*interrupting*) I'm presuming—that you are—*someone.* Doesn't that sound significant?

Client 2: You didn't let me finish my sentence. I'm presuming that you are someone who has done more harm than good by virtue of what you, how you practice your profession for people like me. And that *is* significant.

Therapist D: All right, I just want to gently put before you, I just want to put before you gently now, that . . . that . . . see—if I say to you, "I'm presuming that you are someone"—I knew you were pausing before continuing the sentence, but, you see how that tends to bring out the question as to whether you really are someone or not?

Client 2: No, I don't. What signif . . . why should that . . .

Therapist D: Slips of the tongue or the exploration of slips of the tongue were never of any use to you, hmm?

Client 2: Well, ah, first of all, I don't consider that a slip of the tongue. And . . . and . . .

Therapist D: I don't either, but it's something in the general realm, an opportunity. This interview is an opportunity. I'm suggesting to you that you passed up an opportunity to learn something.

Client 2: If you were actually in the one-up position that psychoanalysts are trained to take, then I guess that would be an opportunity I missed, but I'm not going to let that happen anymore.

Therapist D: This shattering laugh of mine made you somewhat nervous, did it not?

Client 2: I see *you* as somewhat nervous.

Therapist D: You do.

Client 2: Yes.

Therapist D: That's begging the question whether my shattering laugh made you somewhat nervous.

Client 2: No. Don't forget, I've been born, not born, but I've been raised by people like you. My world has been . . .

Therapist D: Raised? How do you spell *raised*?

Client 2: *Bred*—B-R-E-D—as in . . .

Therapist D: You're not familiar with two ways of spelling *raised*? How about R-A-Z-E-D? As in I've been *razed* by people like you.

Client 2: You are being so . . . and so what if it were true . . .

Therapist D: Well, I was, in essence, speculating whether if I were married to you, whether I would be a carpet, or a rug, you know? Dominated, subdued by you.

Client 2: I think you're threatened by me, I think that's why you're saying, "If I were married to you, I'd be so threatened by you, I'd be . . ."

Therapist D: I'd be *razed*. Would I be *razed*?

Client 2: Bullshit.

Therapist D: You can't believe I might be *razed* in that sense?

Client 2: I don't think that's relevant. And no, you wouldn't be razed by me. And I wouldn't be married to you, that's for sure. You're . . . I know you're going to say you hear a lot of anger, yes, there is a lot of anger, because you're bringing up things that are irrelevant. Just like the psychoanalysts always have.

Therapist D: (*speaking over client's voice*) I didn't necessarily hear a lot of anger. I was imagining you saying that to the man you're living with. I was just imagining you saying to him, "I would never be married to you, that's for sure." Does he get traditional, does he?

Client 2: And that *is* bullshit, and I'll tell you that right now, because that's not true.

Therapist D: He's totally free from traditional values in that regard, he doesn't, ah . . .

Client 2: He's a man who has been married three times in three different countries. He has four children. He has had a drinking problem until a few weeks ago, when he, when he started a few months ago also to go to the man that I was starting to see . . .

Therapist D: (*interrupting*) Same one, huh? Hm . . . He is . . .

Client 2: He no longer drinks, he's doing a lot better . . .

Therapist D: He is. And your mother, did you say your mother is also going to this man you go to?

Client 2: Yes. And now my mother is because she has seen changes . . .

Therapist D: (*speaking while client is still speaking*) . . . in, ah, you, in . . .

Client 2: . . . in myself, in my boyfriend, she has never seen any changes in herself, nor my father, and they've been going to somebody else for years now, we are a family of, well, we've all been born and, I shall use the word *bred,* by a psychotherapist, and the only person in my family—and that includes a family of more than myself who have been in a psychiatric hospital and an aunt who committed suicide, and a grandfather who has almost been committed to a psychiatric—I am the only one in my family, and boyfriend, who has made a dramatic change.

THE SECOND INTERACTION

Therapist E: Well, let me share with you what I find myself experiencing at the moment because it may be of some relevance to some pattern in your life, see, ah, irrespective of whether therapy could help or whether different therapists could have helped, whether therapy could have helped, I, if we can put that to the side for one moment, ah, I find myself reacting to you as though, well, ah, my trade has nothing to offer you. As though you're telling me, in effect: "Who needs *you*?" See . . . now, ah, because you're speaking so eloquently and with such fervor about . . . these other ways of getting help and that's done something for you or seems to have done something for you, or you feel it *could* do something for you, whereas these twenty years of therapy is just so much bullshit to be washed down the drain. (I'm exaggerating to make a point.) Ah, now, that, ah, feeling I found in myself of being told I'm useless, ah, it's a feeling of rejection. I wonder how central that is to what you have been struggling with your whole life of being made to feel useless and rejected? Is that part of, is it a great pleasure finally to be able to make somebody else feel that he's useless and rejected? Where the shoe's finally on the other foot, if you see what I mean . . .

Client 2: No, no, ah, no, I think that what it is is a great need to say to the world of psychology that there's—something *missing* . . . and . . . and, see, I had said on the other tape, my aunt committed suicide, three of my best friends committed suicide, because naturally I knew other people like myself. I guess we seek each other out . . . ah, tsk. There's something missing and I don't want it to be going on anymore. I think that people spend so many years and so much money in therapy and all I'm saying is that there's something missing and I want this world of psychiatry and psychology to look into it to find out, because it's only since I've been with the Anonymous groups that I really have felt a lot of change take place, and I am pro . . . I'm still propsychology. Matter of fact, I'm a super . . . I'm under the supervision of my therapist, I'm doing some lay counseling now and I'm really enjoying it. But, ah, ah, and I . . . I'm trying very hard to make sure that I'm providing something, whether it's instructions, even let alone insight, to get people to get out there and do what they need to do to change, because you have to do something in order to experience something in order to internalize something to change and it doesn't really, therapy is such an idealized situation, you feel comfortable with your therapist . . .

THERAPIST E: Therapy, I hear you saying, is for people whose mothers loved them, it's not for people like you.

CLIENT 2: But it is if it—if the world of therapy can find, can hook on to a few things they have never focused on before.

THERAPIST E: Like that—like what I was saying to you a moment ago, which I then had . . .

CLIENT 2: Maybe I didn't understand about mothers . . .

THERAPIST E: . . . the feeling that you were not listening, so I ended up with that instantaneous feeling of uselessness again, see . . .

CLIENT 2: No, well, "mothers who love 'em"—I don't understand what you mean.

THERAPIST E: Well, if I hear you right, from about twenty minutes back, you're indicating that your mother was, at least the way *you* experienced her—never mind how it was, what she would say, what neighbors would say—you felt that she didn't love you in a particular way, that she didn't love you in a, ah, all-accepting kind of a way, and that has been something that you can still try—that has affected you—and that you've been trying to deal with and make up for all your life and that, as I say, therapy—the words and the intellectual understanding and even the emotional understanding, ah, of one's problems—is something for people who have had a little mother's love under their belts, ah, have had—where the therapist and the patient can take it for granted that the patient is not starting from so far back as not even to have had a taste of that, and that for those who haven't tasted it, ah, there is such a sense of lack and such a sense of deficit that even the warm therapist who works with one for a long time can't make it up because that isn't what we offer.

CLIENT 2: Yeah.

THERAPIST E: We do not, ah, become the good mothers that patients didn't have if they had lacked them.

CLIENT 2: Yeah, I can see I must . . . I, ah, I guess I'm, it's funny, I . . . I guess I'm more dedicated to wanting this to be heard than I was to listening to you, because ah, I—it makes per . . . What's—what you're saying now, "the mothers who love 'em," I can see it, now it's all kind of *obvious* what you're saying, so I guess I wasn't listening well enough.

THERAPIST E: To listen might be to become sad.

CLIENT 2: Mmmm?

THERAPIST E: To listen might be to become sad.

CLIENT 2: No, ah, no—I think at this moment in my life I'm having this, more of this need to be heard because, having gone through the twenty years, I'm wanting to protect people who might be going through it now or young children who . . .

THERAPIST E: And about yourself . . .

CLIENT 2: . . . in their twenty years. Oh, yeah, but I'm feeling like I'm working on myself. Like, I feel now everything is probably going to be okay, because I'm beginning to walk the twelve steps, I've got my sponsor, I've got my group of people who have been either been in group before or been through it, you know, I feel like, okay, I found a process that is working for me. What about, for example . . .

THERAPIST E: But isn't that the same as saying that you found a process that when you become absorbed in it, it . . . ahm . . . helps shield you from the sense of this terrible deficit that has plagued you all your life, for two reasons . . .

CLIENT 2: (*sigh*) I don't think so.

THERAPIST E: . . . one, because you talk over it, and two, because you provide other people, other unloved children, as it were, something that—you become the mother, the better one, the one who is there for them and who is friendly and is sweet and who helps them through the steps. You've become the mother you didn't have.

CLIENT 2: Yeah, I do do that, yeah, and . . . and . . . uh-huh, I've been working very hard at not doing that so much anymore.

THERAPIST E: Well, I don't know if it's bad.

CLIENT 2: That yes, but . . .

THERAPIST E: I'm . . . I'm not saying that as if to . . . to comment, to criticize it; I'm saying that, ah, ah, there is a quality of you that talks over this deficit, and I—it comes across to me that you try to blind yourself to it, ah . . .

CLIENT 2: Oh, I try, I try not to feel it.

THERAPIST E: In such a way now as a therapist I would feel, ah, that you were constantly, ahm, not being tuned into me—that you were leaving me to one side, that the relationship was, it would be tilted, one-sided.

CLIENT 2: Mmm.

THERAPIST E: I don't know whether that kind of thing, ah, shows up in other relationships of yours. I would have to ask you about whether there's been some uneasiness in your relationships with your boyfriends, because that would be a problem, where you tend, in order to erase the sense of this ancient pain, to talk, dominate, to overwhelm the other so as not to allow yourself to be vulnerable. I would wonder that—as to how it is with you and those who have been close to you, those that you permit to be close to you.

CLIENT 2: I would say no.

THERAPIST E: No what?

CLIENT 2: That it isn't—that isn't what happens.

THERAPIST E: Oh, well, so much to the good.

CLIENT 2: Oh, (*laughs*) well, my boyfriend and I met in a psychiatric clinic and we've always tried to be mutually supportive and it feels very good that way, and, no, I don't think that I try to dominate people.

THE THIRD INTERACTION

THERAPIST F: Can I interrupt for a moment?

CLIENT 2: Sure.

THERAPIST F: Because I find a curious thing happening, that each time that I try to pursue a line, you go off at a slightly different tangent and bring up another idea, and I find, uh, that whatever line I was trying to pursue doesn't get any further.

CLIENT 2: Mmph.

THERAPIST F: Ahm, I do get one thing clear, that you are very dissatisfied with everything that has been done for you throughout your life—a feeling that you didn't get what you should get, and I would think that might explain why you eat too much, that if you don't get satisfaction, then you're always hungry.

CLIENT 2: Yeah, but what is, well, a symptom stays as long as the disease stays.

THERAPIST F: It keeps repeating over and over again, that's where we started, isn't it? And, of course, if you don't get satisfaction, whether it's a Freudian or a Jungian or a psychoanalyst or a job or a person, I could see that one goes on eating more and more, 'cause one never has a feeling of satisfaction within, so one's always hungry and never satisfied. Now I just wonder, since I've almost had to force you to come to a halt, we could have gone on to the end of this session or maybe many sessions without you feeling you got anything satisfactory from *me,* if you never allowed me even to finish anything. See, I . . . I get the feeling that in some ways you . . . you . . . you were saying that you had the feeling that the therapists are more interested in their points of view than in the patient, I got the feeling that you were more interested in your point of view than in mine.

CLIENT 2: So you're wondering if this is how—this is how I have been in therapy all these years, more starting out . . .

THERAPIST F: Something like that, something like that, whether . . .

CLIENT 2: I've really been dedicated to my therapist's point of view for, one therapist or the other, for twenty years, I've been more of a listener than a talker. This is rather new for me.

THERAPIST F: Well, then, maybe it isn't just listening and talking. Maybe it's something to do with an idea that, ah, what comes from outside isn't likely to satisfy you, that somehow it always fails, whether it's a mother's feeding or a mother's attention or love or whether it's a therapist's interest or whether it's a particular point of view. If you have that feeling that they're always interested in themselves, whether you ever get . . .

CLIENT 2: I never had that feeling until last Monday—that's the first time I ever thought about it.

THERAPIST F: And yet for twenty years there's been this feeling that you didn't get what was needed to make progress.

CLIENT 2: I didn't think—my conclusion is that I personally didn't do what was needed to make progress.

THERAPIST F: Ah.

CLIENT 2: And I . . .

THERAPIST F: So that is a different idea, isn't it?

CLIENT 2: Yeah.

THERAPIST F: See, your first idea was that somehow we . . . we . . .

Client 2: No, it's not a different idea. I don't think that therapists are prepared to help us do what we need to do . . . that's what I'm saying.

Therapist F: I see.

Client 2: I'm saying there is a dual responsibility here.

Therapist F: Uhm, it's interesting because we were, ah, talking about first of all the idea that the therapists were at fault, and then I tried to focus at this moment on was there a part you played and you quickly moved away and said, "Oh, it's dual," almost as though to consider any part that you play may be frightening or painful or that you're reluctant to consider what part you may have played in these situations failing.

Client 2: No, I'm not afraid of that—that's why I stopped the therapy, really, to *think.*

Therapist F: Uh-hum.

Client 2: Ah, I don't know . . . er . . . er . . . you know, I don't know what would be frightening about . . .

Therapist F: Being a—the—*patient.*

Client 2: Being responsible for my—NO, my—you see, I've had a lot of chance to observe: I've seen my aunt commit suicide, I've seen three of my best friends commit suicide and naturally birds of a feather flock together, they were all like me. Ah, my boyfriend is like me, I've seen this therapist charge him forty dollars for five minutes in order to get a new prescription, a new antidepressant that just keeps him where he is. Um. I have had psychiatrists tell me my prognosis is poor because, because I've been this way for so long, I've had most psychiatrists spend most of their time ehr, saying, "Ah-hum, uh-hum, well, what do you think of that? Uh-hum." It wasn't until, and, listen, I do believe that I'm responsible for myself. I also believe that a therapist—if the therapist is seeing a client—is responsible for staying on top of what can be done for a person's problem and I'm only trying to say that I have discovered lately that, and I'm not, I . . . er . . . it doesn't interest me who is more responsible, my interest is, ah . . .

THERAPIST F: But it would be interesting, wouldn't it, and important . . .

CLIENT 2: . . . who is what could be done, what could be done . . .

THERAPIST F: . . . important if there was something that you, a part you were playing, because there's something that could be changed by *you.*

CLIENT 2: But that's why now . . .

THERAPIST F: If . . . you . . . you . . . your first idea is to change all the therapy . . .

CLIENT 2: No, but you see, no, what I'm trying to say is that I think I've discovered something for people like me that would help, it's not new, it's not original, ah, but I started looking at institutions where people were changing, were being helped, people like me, people who are alcoholics, drug addicts, overeaters, not ba . . . barely functioning people, people who have . . . eh . . . find life meaningless, who find suicide seductive, and the one place that I have seen most of these people helped is with the Anonymous organizations and I thought, you know, it's really interesting to see what they do and that it's the people who are taking . . . they . . . they . . . people taking responsibility for their own lives with the support of other people like them and how they go step by step toward being healed—being healed, it's like a wounded healer healing another. I think that my therapist in part wasn't able to help me because she's never been in my position and, ah, ah, it . . . maybe it would have enhanced the therapy if she would, but on the other hand, ah, the woman who . . .

THERAPIST F: You think that . . .

CLIENT 2: . . . helped Sybil was never in Sybil's position either, and she helped Sybil . . .

THERAPIST F: Mmm.

CLIENT 2: . . . so I'm not trying to lay blame on one area or the other—I'm saying it's time to look at problems in a more threefold perspective than psychology, it seems to me that psychology in general does—and I think it would help people more. I'm not laying blame on anybody or anything. I . . . I let I do have some anger toward a lot of, ah, of, ah, whether they be psychologists or psychiatrists who do lay back and don't do very much and, ah, sometimes these people with medicine, I do have some anger for that . . .

THERAPIST F: And that seems to be the main thrust of your feeling.

CLIENT 2: But I don't *think* about that all the time.

COMPARING THE SESSIONS

As with the first trio of cases, I deliberately chose excerpts in which the client recounted some of the same life experiences (for example, the family background of psychiatric illness and suicide, and her discovery of the twelve-step program after twenty years of therapy). But these excerpts are interesting for another reason. They indicate the similarity of the therapists' experience of this client, along with their individual ways of expressing it to her.

Therapist D goaded the client somewhat unpredictably, evidently attempting to get an intense emotional response from her. Therapist E spoke of feeling rejected and useless, trying to show the client that she was putting him in the position she had endured with her mother. Therapist F explained at length the infantile psychology he thought lay behind the client's approach to him. Accordingly, the client's immediate responses to each therapist were quite different. With Therapist D, for example, she was unable to engage in the long intellectualizations she produced with the other two therapists. Therapist F, on the other hand, frequently spoke at length himself, attempting to assert his presence and get her attention.

TESTING THE FIVE DIMENSIONS

Our informal impressions of the consultations are naturally restricted to issues of apparent meaning. We wanted to go deeper, into areas that are not immediately observable. We wanted to know the actual effects of the speaking parties on each other's productions, their influence on each other—at least in terms of storytelling propensities and images.

Our initial efforts, however, were modest. We had to first ascertain whether the five dimensions we'd decided to measure were, in fact, worth measuring. Thus, we framed our initial exploration in global, statistical terms. We asked: Do changes on any of the five dimensions in one speaker correspond in some regular way with changes in these dimensions in the other speaker?

As I explained earlier, the discovery of a statistical relationship can take us only so far. For example, the height of corn can be correlated, up to a point, with the amount of rainfall or fertilizer used; the amount of aggressive imagery on a Rorschach test can be correlated with the degree of a trauma a person experienced in early childhood. It would be a mistake, however, to understand these relationships as causal. Statistics can't tell us how a relationship actually works or comes about.

We used statistical correlations in our initial project because we were testing the viability of our five variables. We wanted to know whether a change in one person corresponded to a change in the other a sufficient number of times to be worth our attention. Our results would tell us whether we could assume a temporally synchronized dance going on between the two speakers. The actual nature of the correspondences was a matter we'd deal with later, with other methods of analysis.

CHAPTER EIGHT

The Psychotherapeutic Dance

If the observer affects the observed in a dance
of mutual interaction, then the question of truth becomes,
"What kind of dance do you want to have?"

—HENRY REED, *PERSPECTIVE*

WHEN WE TALK about the effect that one person has on another, we don't usually think about qualities that are inherently measurable. A person's sensitivity to another involves too many unspecifiable variables—innate disposition, socialization, context, emotional attachment, and so forth.

On the other hand, it's entirely possible to observe that certain kinds of behaviors elicit predictable responses from others. For example, when Fred Astaire dances with Ginger Rogers, we may be struck by how easily they sense one another's cues.

A good football team illustrates the same principle. The players consistently anticipate one another's physical decisions. A good coach will spend a lot of his time analyzing the team's choreography, with an eye toward streamlining his men's responses. He does this by slowing down a film of the team's last game and paying attention to the players' interactions with each other.

Now, suppose, for some reason, we wanted to do the same thing with Astaire and Rogers. Suppose we slowed down one of their films and simply recorded their dance positions from one moment to the next. In our analysis of this data, we might realize that each time the heel of Astaire's hand grazes the small of Rogers's back, she moves her right foot slightly outward, anticipating his lead on the next whirl.

As the Puritans once observed—rather sourly, it's true—time-synchronized dancing is essentially lovemaking raised to the status of art. All manner of natural human interactions operate this way. One person's body language corresponds with another person's change of behavior. In statistics, such a correspondence is called a *cross-correlation.* The behavior of one person correlates with and affects the behavior of the other—usually with some delay or time lag.

It seemed to us that the five dimensions of communication could be treated like this in our consultations—like the physical aspects of a dance. As the amount of strong images, new themes, or positive/negative tone in one person's communications changed, perhaps there would be a corresponding and later change in the other's dimensions. We decided to look for such changes within a ten-minute period of time (five minutes before and five minutes after the target event).

It should be noted that cross-correlation studies are usually undertaken to quantify the effects of a force or energy on a material entity. This is done by noting correlations between two sets of data separated in time. One might say that we carry out this kind of study informally whenever we adjust a flame beneath a pot of spaghetti sauce. We're essentially correlating the changes taking place in the sauce (a liquid entity) with the level of heat (energy) we're applying to it. A comedian does the same thing when he assesses an audience's response to his humor, seeing how long it takes the majority of his listeners to "get" the joke, perhaps adjusting his material accordingly.

We didn't think we were going too far afield to theorize that cross-correlations on our five dimensions might tell us something about the level of emotional or psychic energy activated between a client and a therapist. We had, in fact, chosen our variables because we had reasonable expectations about their significance in this sense.

For one thing, I knew from my analytic experience that a client's stories often communicate, at a secondary level of meaning, unconscious perceptions of the therapist's efforts. But more than this: The speaker duration results had indicated that therapists *react* to such stories—even when they don't recognize their secondary meaning or import. I mentioned this reaction in Chapter Five, with respect to stable speaker systems.

Recall that all our dialogues had a regular and predictable pattern for the switching of the speaker role. Some were more stable than others, but none was chaotic or unpredictable. Thus, pattern breaks, when they occurred, were clear-cut. When we checked the transcripts to see what was happening at the time of the breaks, the situation was always the same. The client was telling a story—usually about a person or a situation outside of therapy—whose images conveyed strong unconscious perceptions of the therapeutic experience. As the intensity of these images increased, the therapist would intervene unexpectedly, both cutting off the story and changing the established speaker pattern.

Because pattern breaks in our dialogues usually clustered around narrative images and interruptions of this sort, Dr. Badalamenti and I theorized that the effect of one speaker's communications on another's might be understood as an observable change in a communicative system driven by increased emotional energy. We did not intend this hypothesis metaphorically. Communicative output is a product of the human mind, which is part of physical nature.

As I said earlier, the products of the human mind are demonstrably material. They expend energy. Our thoughts and feelings occur as the brain consumes glucose and works to create neural connections. By quantifying the narrative images the speakers produced—indicating their extent, thematic novelty, and tone—we hoped to detect the movement of emotional energy as communication changed across a session.

At the very least, we hoped that correlating changes in these dimensions would give us a statistical basis for inferring a deep level of influence between the speaking parties that operates outside of awareness. This influence does not occur in terms of meaning, but in terms of the communicative vehicle. For example, if a therapist consistently increases her use of negatively toned images on average 93 seconds after the client reduces his amount of storytelling, it is not likely that either party will be aware of the correspondence. We thus hypothesized that the greater the size of such cross-correlations, the more sensitive the speaking partners had been to each other—at the level of inner nature, and not as a conscious choice.

One final point: We investigated the behavior of these five items separately as well as in combination. We wanted to know which items accounted for the greatest and least number of effects, and, hence, which dimensions of human communication are most and least open to influence. The sum of the variables determines the system, but some could be more decisive than others, and it was important to know how each behaved in isolation.

CHAPTER NINE

The Skin of a Living Thought

A word is not a crystal, transparent and unchanged, it is the skin of a living thought and may vary greatly in color and content according to the circumstances and the time in which it is used.

—OLIVER WENDELL HOLMES

EARLIER IN THE book, I mentioned the idea of boundary conditions—the fact that a boundary is like the skin of a fruit, or the skin of a body. It houses the system within, but it's also part of the exterior environment, the site of an exchange between the system and what lies outside it. Verbal communication can be understood in the same way—as Oliver Wendell Holmes so nicely puts it, as the skin of living thoughts. Our vehicles of speech represent a site of contact between our neurological system and the outside environment. By assigning values to certain kinds of vehicles (our five dimensions), we hoped to find their relationship to the environments that evoked them as adaptive responses.

This effort is a little like an archaeological dig, in which one determines whether a treasure lies beneath one's feet in a desert or ice tombs in the Asiatic steppes. Such determinations involve a certain degree of intuition and good fortune. They also require noting signs that are visible on the surface of the site.

For example, in the early nineties, a spring drought turned a river in Israel into a mud-soaked plain of debris—wooden fragments mostly, and among them, small coins and arrows surfacing in the shallows. These artifacts were the remains of a Galilean boat, which had been sinking into the river's floor for over two thousand years. Until the waters receded, however, and pieces of its cargo came floating to the surface, whatever signs may have pointed to its existence were too subtle to be noticed.

Essentially, we were trying to isolate in our therapeutic consultations those moments in which deep emotional engagement was intense enough to generate signs on the surface of the exchange—rather like those coins and fragments that became visible in a situation of environmental extremity.

We looked for these communicative signs first, by scoring each of our five dimensions every ten seconds for both client and therapist. Let's take, for example, the lines excerpted in Chapter Four, from the session between Client 1 and Therapist A. On a scale of 0 to 5, the following lines:

> *I was dancing as a little child on the stage, and so forth, and when I would feel I did a good performance, then I felt, I mean, that's what I felt was the only thing in life that made me feel cared about.*

would rate a 5 on extent of narrative, because they contain a strong and definite narrative image. A purely intellectual query, such as the therapist's "Do you feel that?" would rate a 1 on extent of narrative. Just to make sure that the scoring was consensually valid—especially where judgments about tone were concerned—we had the transcripts rated by more than one person, drawn from a group of volunteers who were not involved in the research project itself.

When the series of scores were completed for both parties, we used software to compute the cross-correlations between the series of five items for each speaker. On locating any change in the scores on one item, the computer was to search for changes in the other speakers' dimension scores within a ten-minute block of time (five minutes before and five minutes after the target change). Such cross-correlations are called *CCFs*.

We were looking for a regular change in one person's scores that occurred at the same time as or at some consistent time after a regular change in the other's. Such changes could be positive (increasing or decreasing in the same direction) or negative (moving in opposite directions).

Given two speakers, five dimensions, and a fifty-minute session (of which only small portions were excerpted in Chapters Four and Seven), there were twenty-five CCFs possible across five ten-minute blocks of time, or 75,000 seconds of potential influence. We began the project by looking for cross-correlations on only two of our variables. We chose the ones we thought most likely to influence each other: the client's introduction of new narrative themes and the therapist's output of narrative imagery.

THERAPIST F AND CLIENT 2

We selected our first test case at random—the one with Therapist F and Client 2, excerpted in Chapter Seven—and we ran into a problem almost immediately. Of 75,000 potential cross-correlations on our two dimensions, the computer found none of any significance. Not one. We had been anticipating at least some cross-correlations in all the consultations—on these variables in particular. Had we misjudged the nature of our dimensions? Perhaps there was something odd about the case.

The transcript had seemed straightforward enough. Indeed, on the surface, the consultation with Therapist F had been interactive and engaging. As I said in Chapter Seven, Client 2 was a seriously overweight woman whose stated agenda was to use the interview as a bully pulpit for the twelve-step program she had recently discovered. All three therapists regarded her evangelistic pronouncements as an intellectual defense, but Therapist F was intent on explaining its psychological source, as he understood it.

Throughout the session, he linked the client's compulsive overeating with a deeper hunger for love and attention, undermined by a paradoxical inability to take in anything when it was offered. He believed the client's proselytizing attitude toward him was a good illustration of the phenomenon. She was refusing even to "taste" his viewpoint by insisting she had found something better and didn't need his help.

Therapist F is an experienced senior analyst, and his interpretation of the client's defenses was both cogent and perceptive. Its surface meaning, however, was only part of what the client was experiencing in the interview. She produced several narrative images that suggest strong unconscious perceptions of the therapist's efforts. For example, in her contrast of Overeaters Anonymous with standard psychotherapy, she says at another point in her session:

> *All of us [in the OA group] become—it is a give-and-take situation with some definite guidelines. That is so fascinating to me—the fact that action has never been emphasized enough in therapy, mostly sitting and talking and thinking and gaining insight and theorizing and discovering, which is fascinating—you know, it's almost a form of philosophizing, or—it's fascinating, but it doesn't do the trick . . .*

The client's generalization about traditional therapy—as inactive talking—encodes her valid unconscious perceptions of this particular therapeutic environment. Throughout the session, Therapist F did "sit, talk, theorize, and philosophize" at length. In fact, the first image in the client's contrast between therapies is what I generally call an unconscious attempt at rectification. It implicitly advises the therapist that he is not providing the structure the client needs for a genuine give-and-take exchange.

So the lack of deep effects in our test run was not due to an absence of meaningful narrative images. We decided to run another case so we'd have a point of comparison.

THERAPIST E AND CLIENT 2

For consistency's sake, we chose another interview with the same client—the one conducted by Therapist E. This therapist was also interpretive, but he was more inclined than Therapist F to encourage the client to talk about her life experiences and memories. And this time we got a significant negative cross-correlation. Every time the client's use of new narrative images decreased or increased, within ninety seconds later on average, the therapist's use of narrative images moved in the opposite direction.

Because the client's change of score always preceded the therapist's, we credited influence to the client. Whenever the client intensified her use of new images, the therapist intellectualized. Conversely, when the client veered toward intellectualization herself, the therapist introduced narratives into the dialogue. The therapist appears to have been keeping unconscious meaning at a manageable level, albeit without direct awareness. This is the pas de deux we had anticipated—a temporal dance of encoded messages, reflected in coordinated shifts into and out of narrative communication.

Of course, the fact that one result had coincided with our expectations did not confirm the legitimacy of our assumptions. Maybe *this* case was an anomaly. We were, however, reassured by the finding. When we ran the remaining cross-correlational possibilities with Therapist E, we found, in all, 452 seconds of effect. More than half of these changes involved the two variables just described, and nearly all were close enough in time to suggest immediate influence. Moreover, impact went in both directions. That is, neither client nor therapist was responsible for a preponderance of effects on the other.

It should be admitted that 452 seconds out of a potential 75,000 suggests interactional engagement of no great magnitude. In fact, it is less than could be expected on a random basis—so much less that the result was significant for that reason alone. On the other hand, the results both supported and extended the conclusions we had drawn from the speaker duration models. They not only confirmed our observation that an increase of narrative imagery from one speaker can prompt the other to modify the increase by interrupting, but they also established the converse. Therapist E introduced images into the dialogue whenever the client's use of imagery decreased.

This finding suggests, among other things, that speaker influence is not just a matter of which event comes first. Technically, the therapist was influenced by the client's manner of expression, but he was ultimately the one who was determining the nature and extent of images in play.

THERAPIST D AND CLIENT 2

The third consultation involving Client 2 (also excerpted in Chapter Seven) was unlike the other two—at least on the surface. As may be recalled, Therapist D was one of two therapists in the series (the other was Therapist A) who had effectively reversed roles with the client. He interrupted the client unexpectedly and often, but his penchant for holding the floor, when he could, gave the system a certain degree of stability over time.

His style, as I suggested earlier, may have been strategic—an attempt to break through the client's defensive one-up posture. However, the result was a peculiar and rather volatile exchange. The questions he asked seemed to come out of nowhere. They were provocative, highly personal, bordering on the inappropriate, and unusually self-revealing. At first, the client reacted blandly to his provocations, but ultimately, she became both incredulous and confrontational. For example, citing a fresh part of the session:

CLIENT 2: Uh . . . when I last spoke to my former therapist, the one who asked me to do this interview over a year ago, I said, "You know . . . that my position has changed. I don't believe in traditional therapy anymore." And she said that she wants . . . she wanted me here anyhow or we . . . wanted . . . me here as who I am now. So I . . .

THERAPIST D: *She* wanted me here or *we* wanted me here because . . .

CLIENT 2: She used the word *we.* "We want you as you are now." That's what she said to me.

THERAPIST D: Did she make it sound *extremely* sexual? You make it sound sexual to me.

CLIENT 2: You are incredible! You are just incredible!

THERAPIST D: It was the furthest thing from your mind, wasn't it?

CLIENT 2: Sex! I mean, you've got to be kidding!

THERAPIST D: So we want you. You said, "We want you." Is sex so traditional . . .

CLIENT 2: (*interrupts*) I'm almost embarrassed for you!

Based on the jolting qualities of the surface interaction, we postulated that there would be a strong measure of influence here. However, when we ran the twenty-five possible correlations, we obtained a mere 38 seconds of significant effect—nearly all of it involving the negative tone of the speakers' imagery.

A COMPARISON OF THE THREE CASES

Before we had run these cases for cross-correlational effects, we had seen little more than three different styles of interviewing. But once alerted to the presence or absence of interactional influence, we looked at the transcripts anew.

THERAPISTS F AND E

Therapist F, whose consultation had shown no effects at all, had, in point of fact, elicited almost no details about the client's life history or experiences. One might say that he didn't ask, and she didn't tell. The exchange had been so lively that we hadn't realized it was almost exclusively intellectual in nature. But now it was clear that the therapist had not encouraged any stories about events or people. Rather, he had solicited speculations, analyses, conjectures, and comments about theory.

Occasionally, when the therapist became directly critical of the client, she shifted from argumentation to telling a story about someone who had been violent and destructive. But this shift was neither consistent nor dramatic enough for the computer to pick it up as a significant effect.

We hypothesized, for the moment, that deep communicative influence—the temporal dance that goes on between speakers outside conscious awareness—requires a certain level of unconsciously meaningful narrative material: descriptions of events, memories, storytelling, and the like.

Therapist E, for example, whose interview had shown a midrange level of effects, had led the client to explore significant life experiences—in particular, a suicide gesture prompted by the extended absence of her former therapist. The attempt had resulted in her parents' committing her to a psychiatric hospital four years earlier. Therapist E not only pushed the client to talk about her parents' reaction, but helped her to consider the possibility that she had staged the event in order to prove to herself that someone cared enough to rescue her.

It might be noted that in the interview with Therapist F, the client alluded to suicide and hospitalization many times, but, unlike Therapist E, he did not pursue her allusions. Recall the client's statement cited in Chapter Seven:

> *I've had a lot of chances to observe: I've seen my aunt commit suicide, I've seen three of my best friends commit suicide and naturally birds of a feather flock together, they were all like me . . . [but] . . . most psychiatrists spend most of their time ehr, saying, "Ah-hum, uh-hum, well, what do you think of that? Uh-hum."*

Here, the client all but requests that the therapist ask her about this issue. She then follows the request with a clear narrative image—characteristically, about analysts in general—that encodes her unconscious picture of the therapist's disengagement from her.

THERAPIST D

Given the theory we'd tentatively formulated—about an intellectual versus a narrative-seeking approach—the matter of Therapist D fascinated us. As I said, we had anticipated a high level of cross-correlations in this case, because the interaction of the principals had been so heated and powerful, even wrenching. But the little influence we found was concentrated in the area of negatively toned images.

When we looked at the data more closely, we saw that most of the client's narratives had actually been quite positive in tone, but not one of them had any effect on the therapist's communications. Indeed, these positive images appear to have been an unconscious attempt on the client's part to deny her perceptions of the therapist's assaultive nature—that is, to insulate her from their impact. The therapist, for his part, was marching to his own particular drummer, with little awareness of the client's rhythm or needs.

One can see this quite clearly in the exchanges the two had about the suicide attempt just discussed. Unlike Therapist F, who simply ignored the subject, Therapist D apparently recognized its significance, but declined to pursue it directly—despite the client's active protests. Consider the following new excerpt:

Therapist D: What . . . what . . . uh led to your being in the hospital . . . four years ago?
Client 2: Severe depression and, uh, an attempt at suicide.
Therapist D: Mmm. See, if I ask you how you attempted it, I would be so grossly unfair that I won't ask you.
Client 2: I consider that condescending.
Therapist D: Yeah?
Client 2: I mean, you were saying . . .
Therapist D: Yeah?
Client 2: But why don't you tell . . . why don't you give . . . *be* grossly unfair. What do I care?
Therapist D: No, I'm wondering what this is about, that you make me feel that if I am to gain any kind of equal footing with you, I've got to be massively unfair.
Client 2: Then *be* massively unfair.
Therapist D: Yeah, but that's . . . that . . . I don't like doing that.

Then, about four minutes before the session was to end, the following dialogue took place:

Therapist D: You were seeing then the traditional psychotherapist who . . . to whom you made a commitment to do this interview a year ago? Four years back, were you seeing her?
Client 2: I had seen her for nearly eight years.
Therapist D: And you still keep in touch with her, mmm?
Client 2: Mmm-hmm.

THERAPIST D: Well . . . would she kill herself if you stopped keeping in touch with her? How much do you mean to her?

CLIENT 2: No, she wouldn't kill herself and I don't understand the question. I mean, I don't understand why you asked the question. She cares about her patients. I liked that about her . . .

Clearly, Therapist D had already inferred, without directly eliciting the information from the client, that her suicide attempt was related to the absence of her former therapist. Indeed, without the specific facts in evidence, the paradoxical question he asks—"Would she kill herself if you stopped keeping in touch with her?"—while it may be a reflection of his own despair over the end of the consultation, is extraordinarily perceptive. Therapist D's approach has told him a great deal more than he has let on.

But this is exactly what we were seeing in the results of our cross-correlation run. As this therapist marched to his own inner rhythms, he was accumulating insights about the client, but the client reacted by protecting herself against his emotional onslaughts. Her description of her former therapist as someone who cares about her patients is one of those positive narrative images that appears to deny the implicit cruelty of Therapist D's manner of questioning.

It might be suggested, in this regard, that startling someone into unwitting surface revelations has little to do with affecting the person at a deep unconscious level. Some aspects of communication belong to the realm of nature, our psychobiological nature, and must be respected, because they come into play without our volition or knowing.

CHAPTER TEN

Effects or F/X?

I sometimes hold it half a sin
To put in words the grief I feel;
For words, like Nature, half reveal
And half conceal the Soul within.

—ALFRED, LORD TENNYSON, *IN MEMORIAM, PROLOGUE*

THE TERM F/X is popular shorthand for what are generally called special effects—sounds or visuals produced artificially and added to a film or TV show at the processing stage. As we ran the second trio of interviews for our cross-correlation study, we were beginning to make a distinction between natural and produced effects in a dialogue.

For example, a speaker like Therapist D may deliberately unsettle a client in order to elicit an emotional response. This is what we might call communicative F/X—a consciously determined strategy of influence. But there are also unconscious, involuntary effects, apparently produced by natural strategies of communicative adaptation. Although they occur simultaneously, the two levels do not necessarily work in concert, and may in fact be in conflict with each other. We kept this distinction in mind, because we were now attempting to predict, given what we had seen in the other cases, what kind of involuntary effects we were likely to find in the next trio of interviews.

The three sessions we were considering, excerpted mainly in Chapter Four, were conducted with Client 1 by Therapists A, B, and C, in that order. The client was an intermittently successful insurance broker who suffered from depression, alcoholism, poor self-esteem, and an inability to stay with anything for very long. She had lived with several men, and her present marriage, her fifth, was going badly. All three of her therapists elicited a basic life history and pertinent information about her reasons for seeking help, but their styles and her responses to them were quite distinct.

We began, as we had with the other client, by selecting the last interview of the three—the one conducted by Therapist C.

THERAPIST C AND CLIENT 1

This was a session that seemed resonant and productive in both the video and the transcripts. The therapist was highly compassionate and probing, and the client frequently spoke through tears. Interestingly, however, her statements were rarely expressions of feeling. They were largely analytical—observations about her current state of mind. For example, at one point in the session the following exchange took place:

Therapist C: So the idea of being guilty about it, or fearing success, that is something you say you've heard, but that's something you can't easily apply to yourself.

Client 1: I really don't know if it's true or not. It's been brought to my attention that maybe I have that syndrome or something, or—quirk.

Therapist C: You know, that interests me.

Client 1: Oh?

Therapist C: You know, if you feel it, what would make . . . er . . . since you described your childhood as so difficult, here it is, you know, you have a chance to overcome it, transcend it, and yet you say you . . .

Client 1: Right.

Therapist C: . . . feel guilty in doing that. I'd like to understand what might be in your way.

Client 1: I don't know. I'd like to understand, too, why I can't break through that barrier. Mmmm. I have a compulsive-obsessive nature, I think.

Perhaps more significantly, this was the only interview in which a client ended the session by telling the therapist how much she'd enjoyed their time together. As I said in the last chapter, highly positive statements can sometimes be a form of denial—the client's way of protecting herself against an emotionally assaultive technique. Therapist C, however, seemed very much as the client described him: friendly, empathic, and concerned about her feelings. Moreover, the client clearly felt comfortable with him. At several points in the interview, she introduced highly charged material that had not come up with the other two therapists.

Toward the end of the session, the therapist pressured her to talk about her first two interviews. He wanted to know whether she'd gotten anything out of them. The question was inappropriate and provocative, and the client, rather predictably, was led to compare the two previous therapists unfavorably with this one. It struck us that both Therapist C and the client were emotionally engaged in this interview, and we speculated that this mutual attunement would be reflected in our measures of communication and influence.

Thus, we predicted a moderate to strong degree of cross-correlations. To our surprise, the effects were limited to a mere 23 seconds. This was an even lower number than the one we'd seen with Therapist D, who had deliberately baited his client and appeared to trigger her defenses. We were beginning to lose heart. Three out of four consultations had come up near-empty on our measure of deep influence—with little to suggest a common reason.

On the other hand, the very infrequency of the cross-correlations, as I said earlier, was highly significant. They were a good deal less frequent than chance would allow. In a positive sense, the ones we were finding were consistent from case to case. There was little time lag between the correspondences, they indicated influence from both speakers, and they usually involved some activity around new themes and narrative output. So we didn't want to abandon our five dimensions just yet.

Instead, we went back to our speaker duration models—to see whether the case with Therapist C had evidenced the kind of pattern breaks we'd come to associate with unconscious interaction in dialogues. As it turned out, this case had been the least stable of the ten. Now we wondered why. Could the absence of cross-correlations be attributed to the hidden instability of the system as a whole?

Instability, remember, in terms of our research, means that a regular rhythm of speaker alternation is poorly established, or it breaks down too often to be easily predictable. We wondered if there might not be a direct relationship between an unstable speaker structure and a dialogue that won't support communicative influence. No matter how attuned the speakers appeared to be on the surface of the exchange, perhaps they weren't influencing each other at all at a deeper level.

Ultimately, our attempt to clarify what we were seeing led to a surprising bit of information. It turned out that Therapist C had, in essence, established a prior relationship with this client. The two had engaged in a long conversation while waiting for the video crew to set up. Strictly speaking, there were no rules against this sort of thing, but it became clear now that the therapist had seemed emotionally engaged in the interview because he, in fact, was overengaged. The session began, as it were, in the middle. It was a continuation of the discussion that had been taking place before the taping began.

In the course of this earlier discussion, the therapist had encouraged the client to talk about her previous interviews. She said she'd been particularly discomfited by the session with Therapist B. He had asked few questions and forced her to "do all the work." She was hoping, she said, that Therapist C would appreciate her feelings and take the lead in their upcoming consultation. The following fresh exchange, which occurred late in the actual session, is clearly based on Therapist C's having attempted to comply with her wishes:

THERAPIST C: At the beginning, you know, you seemed tentative. For example, you wanted me to ask questions, rather than being forced. Perhaps you might say something about your anticipations and feelings as you entered this interview. What were you expecting from our conversation?

CLIENT 1: Oh, about the same as the other times I was interviewed here. Ummm. I think the first time I cried, the second time I didn't.

THERAPIST C: Well, what did you take away from the first two?

CLIENT 1: Not much.

THERAPIST C: What prevented you from being able to take something away from . . .

CLIENT 1: I don't think the other people really told me, I didn't get any insights, or it wasn't like a regular therapy session. In other words, I didn't, in other words, I was just talking like a new patient in therapy. They have to talk a couple of times before the psychiatrist would know enough about them really . . .

THERAPIST C: Um-hum, so you felt that only you were giving and that you weren't getting anything in a way, and you were afraid that was going to be repeated this time again . .

Our discovery cast this session in a completely different light. Like Therapist D, Therapist C had come into the session with a premeditated agenda. Here was a perfect example of F/X versus effects. What was intended as empathic guidance on the surface of the dialogue wasn't really empathetic at the level of deep nature.

Indeed, given the cross-correlational results, one might speculate that, at the deeper level of nature, it doesn't matter whether a premeditated agenda is manifestly assaultive or accommodating. Either way, it interferes with the deeper communicative interaction that happens naturally as people perceive and adapt to each other unconsciously.

THERAPIST A AND CLIENT 1

Therapist A had been the most active of the six consultants in our test run, and the speaker alternation pattern in her interview had been among the least stable. Both therapist and client spoke in short bursts of dialogue, and the client, accordingly, had little opportunity to tell stories. So, this time, given our developing ideas about the cross-correlations, we predicted a low-to-nonexistent level of deep influence.

Again, our guess was wrong. The session showed a total of 236 seconds of effect. In terms of total amount of influence time, this result was the second-largest of our six. Moreover, the effects were not concentrated in newness of themes or output of narrative imagery, as we'd come to expect, but in negative (and occasionally positive) tone. Indeed, there were few narrative images to speak of.

When we looked at the transcripts, we realized that the dialogue consisted almost entirely of negatively toned *intellectualizations.* Each time the client had attempted to tell a story, the therapist had intervened, substituting a strong intellectual image for the client's narrative one. For example, to cite some fresh material:

THERAPIST A: You don't quite know how to be a good mother to yourself. Maybe that's partly why you're with your psychiatrist now, trying to learn to be a better parent to yourself, do you think?

CLIENT 1: I don't know.

THERAPIST A: To look after that sad and embittered child a little bit.

CLIENT 1: People would, you know, I have been questioned at times, what do I do for pleasure or fun, and I really . . .

THERAPIST A: You don't let that child play very much. We've all got to keep in touch with the childlike part of ourselves, because this is part of the poetry and fun of life. It can be a bit crazy, the child part, and capricious, but it's what makes life fun. You're so very grown up. But if you . . .

CLIENT 1: I think I'm too serious, I don't seem to have . . .

THERAPIST A: Perhaps, yes. Perhaps you're too caring for everyone else, and not giving enough care to the child part of yourself. Not listening to her, comforting her, and giving her time off from caring for the others, and saying, "Look, they've had their caring. Now it's our turn. This is what I would like to do for me."

CLIENT 1: Yeah . . .

THERAPIST A: Do you do that very often?

CLIENT 1: No, I guess I don't know what to do. I've been studying for a year, and, ummm, my husband seems . . .

THERAPIST A: What have you been studying?

CLIENT 1: . . . to think that that's pure pleasure.

THERAPIST A: Well, how good if it were! You need a little of that.

CLIENT 1: I feel that it's sort of a necessity at this point in my life.

THERAPIST A: Yes, studying is hard work, too.

CLIENT 1: And to change, so I can get a better position, and support the family, which he doesn't do.

THERAPIST A: I see. So again you're being the caretaking parent, both father and mother.

CLIENT 1: I guess. I don't know.

Notice how little the client actually adds to this exchange. Each time she attempts to take a narrative direction, the therapist blocks her and takes over the discussion. One might suggest that the therapist was uncomfortable with unconscious meaning and sought to deflect the client's inclination to tell stories of any kind. The client, however, appears to have adapted communicatively, as best as she could, to whatever emotional nuance existed for her in the intellectualized images the therapist used.

THERAPIST B AND CLIENT 1

This last consultation is the one the client described to Therapist C, in their preinterview discussion, as uncomfortable. Having been led by conference organizers to expect new insights into her problems, she had not anticipated this therapist's relative silence, which forced her to do most of the talking. On the other hand, almost all of the talking she did was narrative in nature—descriptions and memories of events and experiences in her life. Moreover, when the therapist intervened, he used those images, attempting to relate what she'd said to her evident discomfort with the interview. So we warily predicted a high number of cross-correlated effects.

This time we were right—but well beyond our expectations. The total number of seconds of significant influence was 2,247—about five times more than the highest of the other consultations. As with those other consultations, the effects ran in both directions (that is, to and from the client), but this case was unique in that the results involved every possible combination of dimensions.

The analyst in this consultation was, as we said, far more silent than the others. He did, however, ask questions and encourage the client to keep going when she fell silent or told him she didn't know what else to say. When he interpreted, he referred directly to the client's images and attempted to tie them to her experience of the session itself. This is one reason the cross-correlated effects were so high. The therapist's responses were guided, even on the surface of the exchange, by the images the client produced. The following new excerpt is somewhat lengthy, but it illustrates Therapist B's distinctive rhythm of silence and speech:

Therapist B: You go back to my helping. Let me ask you again, uh, well, I know very little about how you got here. Could you tell me how this interview was arranged? Could we just sort of start with that? How was it arranged for you?

Client 1: This particular one? Or my seeing a psychiatrist as such?

Therapist B: Well, wherever it begins.

Client 1: Where it begins? (*chuckling*) Well, some years ago, my internist gave up on me, I think, and sent me to a psychiatrist . . .

Therapist B: Um-hum.

Client 1: . . . um, that I saw for several years. Ah, he felt that some of my illnesses were brought on by my mind in some respects . . .

Therapist B: Um-hum.

Client 1: Uh . . . I went to a stress management clinic here in the city and, um, I've been through many other types of things that were supposed to help me . . .

Therapist B: Um-hum.

Client 1: Because of the poisoning, perhaps, they thought that the pains would come on from something in the brain that triggered a remembered response, an anxiety, or a virus that would attack me, would trigger this chest pain. And I really don't know if that's true or not . . . you know, I *really* don't know.

Therapist B: Um.

Client 1: I do have some viruses that reoccur, about three of them, that keep coming back. And it seems at times that a viral episode will trigger an inflammation . . . I don't know whether there's a weakness there or it's something remembered, so I don't know how to guard against it. You know, if it's mental, is there a way to guard against something triggering it by emotional upset? I'm in the process right now of perhaps leaving my husband, and I'm not really all that upset about it at the moment. But I don't know, you know. If and when the actual move comes about and the things, the physical things start moving out . . . you know, then I really don't know . . .

THERAPIST B: Um.

CLIENT 1: I would like to get to the place where I don't react unduly, or overreact. Sometimes I'll get caught up in my mind and write scenarios of what is going to happen. You know, bad things, all the what-ifs—if such and such happens. I seem to be oversensitive in this respect, but a lot of the alcoholics I meet at AA seem to be sensitive this way—you know, taking chemicals in order to cope with an imperfect world, or their imperfect selves, or the people around them. The coping mechanism seems to be a little broken. How do you feel about that? Are we more sensitive?

THERAPIST B: Well, as human beings, we're often—and I think you were sort of hinting at this—we're not in touch with exactly what's setting us off.

CLIENT 1: Yeah . . . right.

THERAPIST B: I do notice very dramatically that whenever I start asking you about how you got here, why you're here, how the interview was arranged, you talk about unpleasant things—terrible scenarios, like poison, pain, viruses, leaving your husband. I think I was moving you into an area where you have some discomfort—actually, I think you feel I'm pushing you into an area of discomfort . . .

It should be recalled that Therapist B's session was the most stable of the ten. There were virtually no interruptions, no struggle between the speakers to take and hold the floor. One can see, from the excerpt just quoted, how the client was led by the therapist's silence to produce longer and longer speech events—many of them narrative in nature—until she actually interrupted herself to ask the therapist a question. Consciously, the client may have felt pressured, but at the level of unconscious communicative adaptation, the interview had a great deal of impact.

It seemed to us that the results on this case supported some of the conclusions we had drawn from our findings on the first five. A stable speaker system appears to support a high level of narrative imagery, which, in turn, provides more opportunity for unconscious expression and communicative adaptation to occur between speakers. More simply put, allowing a person to talk for long periods of time, without preconceptions, appears to have deeper emotional impact than direct emotional persuasion at the surface of a dialogue.

SOME LINKED FINDINGS

We now had a basis for comparing our six cases. We were particularly interested in the two that had obtained the highest level of cross-correlations: Therapist E and Client 2, and Therapist B and Client 1. Therapist E, remember, was the consultant who had led Client 2 into a long discussion about her suicide attempt and the subsequent reaction of her parents.

These two therapists had conducted the most stable of our ten interviews. They were least likely to interrupt their clients, and the clients, accordingly, had shown the greatest inclination to take the speaker role and to continue speaking. (These properties were quantified by low Poisson values—indicating little interruption and continuity of inertia.) Both therapists had evoked a large amount of stories and new themes, but apparently for different reasons. Therapist B did so by remaining silent and focusing on the client's imagery when he did speak. Therapist E pushed his client to recount incidents and memories.

Therapists A and D had created the least stable of our ten consultations. In both cases, communicative impact was limited almost entirely to negative tone. Therapist D had attempted to startle his client into a direct emotional response, but his strategy had little influence at the unconscious communicative level—at least in terms of our CCF measurements. The system created by Therapist A obtained a somewhat higher number of cross-correlations, as suggested, because of the client's unconscious response to some of the therapist's intellectualized images.

A FOLLOW-UP STUDY

In order to further clarify our results on the six consultations, we programmed the computer to locate cross-correlations between our five dimensions and the kinds of comments made by the client and therapist. That is, we opened the same ten-minute window and looked for correspondences between our variables and the way a communicative vehicle was structured.

We defined ten specific kinds of communication—a question, a clarification, an explanation, an interpretation (identifying an unconscious element), a piece of advice, an unsolicited self-revelation, a monosyllabic comment (like uh-huh), a reference to childhood, a direct allusion to the consultation or its conditions, and a description of an event. We wanted to know whether a speaker's selection of a particular kind of comment corresponded to any change in the other's use of tone or imagery.

The results we got were strikingly similar to the findings we already had. The fewest cross-correlational effects were obtained with Therapist D, who used unpredictable questions and self-revelations to provoke reactions from his client (34 seconds in this second study). The interview with Therapist C, the analyst who complied with the client's request for empathetic questioning, obtained 149 seconds. Therapist F, who sought to explain the psychological nature of the client's eating disorder, created a system that showed 193 seconds of effect.

The highest values were again obtained by Therapists B (3,004 seconds) and E (1,126), both of whom, as noted, had encouraged or allowed the client to tell stories. The interview conducted by Therapist A obtained 512 seconds, again, largely on the basis of its negative tone.

As with our first cross-correlational study, apart from Therapist A's negatively toned intellectualizations, most of the correspondences involved the dimensions of new themes and degree of narration. These two variables accounted for 60 percent of the influence in the first run, and for 56 percent in this one.

It appears, then, that our unconscious attempts to adapt to one another occur not only with respect to the attributes of our words, such as tone and choice of image, but with respect to the kinds of comments each party makes.

PARALLEL RUNS

We used our nontherapeutic couples' dialogues as a kind of control group in this cross-correlations project, and we got results comparable to the ones we had obtained with the consultations. The dimensions for each speaker showed a moderate degree of mutual influence that was proportional to the amount of narrative material in the dialogue and to the stability of the speaker duration patterns.

In addition, because Therapist B's approach had produced such strong results on both cross-correlation models, we ran studies on two other interviews conducted by this therapist. We also ran scores from a session conducted by a colleague with the same analytic point of view. The total number of seconds of effect were, respectively, 1,660 and 3,110, and 1,846. Again, the stability of speaker duration was high, and newness of themes and amount of narration carried most of the weight of the correspondences.

CONCLUDING COMMENTS

Our five variables had yielded results that allowed for sensible and predictable discriminations among different speaker systems. Moreover, these discriminations had extended our understanding of the stability measurements we'd already taken. The regular alternation of speakers in a dialogue, combined with long speech events, appears to encourage the production of narrative imagery; and the more encoded images there are in an exchange, the greater the cross-correlational effects.

Conversely, a low use of narrative vehicles—whether this occurs because a speaker interrupts, intellectualizes, or consciously directs the output of the other—appears to be associated with an unstable interaction, and deeper effects are weak or nonexistent. This seems to be the case even when there are intense emotional responses at the surface of the dialogue. When the natural flow of narrative images is thwarted, deeper emotional connections appear to be limited or even blocked.

I pointed out in Chapter Eight that cross-correlation studies are usually undertaken to quantify the time-lagged effects of a factor on the workings of a system. Let's say, for example, that we wanted to determine the effects of the sun's heat (a form of energy) on an ecological system. One of our consistent findings would be that changes in temperature are not as great on islands or along seacoasts. This is because water stores heat. When the sun goes down, the stored heat is released to the environment, keeping the temperatures higher. The correspondences that would suggest this chain of events are time-lagged. Absorption and release occur hours apart.

We hoped that the cross-correlations on our five dimensions would tell us something similar about the time-lagged effects of emotional or psychic energy activated in a client/therapist system. Communication is, as I said, a product of nature, and, as such, it requires a physical expenditure of energy. Our studies had indicated that such energy can affect a person's dialogue.

Accordingly, we theorized that the correspondences we were finding were a way of observing the flow of emotional energy across a session. Time-lagged effects argue for a therapeutic situation like the ecological one I described. Certain therapist behaviors appear to precipitate subsequent narrative images. The time lag between one event and the other is like the lag between hearing a joke and "getting" it. We thus took our research in a direction that would tell us whether this was actually happening and, if so, how it was taking place.

CHAPTER ELEVEN

Catching the Wave

Speech is power.

—Ralph Waldo Emerson

So far, we had established a number of things about dialogues. In each of our consultations, the patterns of talking and listening were strikingly individual, unique to the partnership involved; yet all were stable and regular. Even the least predictable patterns were in no way purely random or chaotic. Over the course of a conversation, each pair brought the lengths of their contributions into harmony. And the speakers' influence on each other was consistently reflected in the way they used images—their number, tone, variability, novelty. Far outside conscious awareness, each participant was adapting to an evolving, jointly created system.

As living organisms, of course, we are always adapting to our circumstances. Although we may not recognize shifts in narrative imagery as adjustments of this sort, communicative responses appear to be part of a continuum that includes our more obvious physical reactions to the environment—particularly the ones that we associate with emotion.

For example, we know from experience that a sense of danger can quicken our pulse and shorten our breath. Such reactions are not under our conscious control. They come from simple perception and inner brain response. The sensory apprehension of a threat—a loud noise, a sudden movement—will orchestrate a whole body response, flooding us with glucose and adrenaline, preparing us to fight or to run. Our first conscious sense of this process is an emotional state. We're roused to action. We feel angry or afraid or excited.

Such involuntary responses belong to a primitive arsenal of coping strategies that are called into play automatically in the presence of the right stimulus. Millions of years ago, these reactions were our primary means of dealing with the environment. Although we gradually evolved the capacity for cognitive decision-making, our more archaic system is still with us. It kicks in too fast for us to appreciate it consciously—at least as the process is happening. The simple act of looking at a clock, wondering if we'll be late to work, can put adrenaline into the bloodstream, just as though we'd sighted tigers on the old savannah.

As stated earlier, our emotionally driven responses are both limited and universal. This makes them easy to interpret when they're visible. A blush, for example, is direct and unequivocal, more basic than words. Thus, we regard the behavior as an unwitting show of feeling. We associate it with self-consciousness or shame and attempt to connect it to a specific external event. To put this another way, we experience such physical responses as a form of meaningful communication.

Other aspects of emotional adaptation occur without observable physical signs. We may not even be aware of them. But they happen all the same—and our bodies sometimes register their cumulative effects, perhaps in high blood pressure or tension headaches.

Communicative shifts in a dialogue are like these invisible physical responses. They're not immediately apparent, and we are not aware of them as they happen. But they do happen. And, judging by our cross-correlations, they have measurable effects for each dialogue. For example, each of our two clients told the same basic stories to all three of her therapists, but the scores on their five dimensions indicate very different unconscious assessments of the input received in each session.

Although the communicative effects we're talking about are largely found by first quantifying and measuring, once they become apparent, they can be treated like any other psychological cue. That is, they can be traced to "trigger" events and recognized as indications of attachment or defense. They, too, contain information on how we effect one another psychologically.

TURNING TO PHYSICS

Now that we knew we were dealing with real, measurable phenomena—events that we could predict over intervals of time—we could move past psychological interpretation to another area of research. The science of physics would permit us to treat our five communicative dimensions literally as entities affected by emotional energy.

We did not intend this venture as an exercise in metaphor. Even though matter and energy are equivalent and can be changed into one another, in classical physics, energy is not considered material. Electricity, light, sound, heat—none of these things has its own space or has weight. We know they exist because they produce measurable changes in matter. This is the situation we were dealing with in our research on communication. We were seeing changes in communicative vehicles that occurred between people and could be anticipated under specific conditions.

Physics has a different eye for each of the forms energy

takes (kinetic, potential, electrical, magnetic, chemical, gravitational, and so forth). The fundamental task in all of these fields is to find appropriate ways of measuring the energy in order to determine its nature and consequences. Once a theory is generated from such measurements, the work passes into the hands of engineers and scientists, who search for ways to apply the theory and solve problems with it.

In many cases, the application comes from recognizing a theory's implications and then implementing them. This is what we wanted to do with our five dimensions. We had set up cross-correlation studies that showed us the effects of emotional energy on communicative expression in a somewhat indirect way. Now we hoped to follow the implications of our results by borrowing models from well-evolved branches of physical science. Part of our research process was to use our data in models developed to measure other kinds of events affected by energy. If nothing else, this would tell us whether the time course of communication was similar in form to the way other kinds of natural phenomena evolved.

BORROWING FROM ELECTRIC ENGINEERING

In the early twentieth century, a conceptual breakthrough occurred in the field of economics. It became clear that market trends could be analyzed and forecast with methods originally developed to measure the strength of electrical signals. The methods worked chiefly because fluctuations in the market resemble the cycles of an alternating current. They peak and fall in recurrent patterns.

The same methods can be used to measure any signal that oscillates in a predictable way—the light from distant stars, the shock waves from a volcano, even electrical activity in the brain. It seemed to us that this model would also apply to the fluctuations of our five dimensions of human communication.

Recall that we chose these dimensions to record and trace

the presence of and changes in narrative imagery across a consultation. We had selected them for two reasons: (1) Narrative images are a measurable component of communication, and (2) they are expressions of unconscious response. We theorized that by treating their fluctuations like waves of electrical current, we might be able to measure the power of emotional energy in a therapeutic exchange.

An electric current is basically billions upon billions of electrically charged atoms in motion. Materials can be used to conduct this flow in one particular direction, somewhat like water through a pipe. We call this one-directional flow of electrical energy DC, or direct current. Its strength is measured in the same way as the flow of water is measured—as the amount that travels between two points within a certain amount of time.

AC, or alternating current, is different. It is generated by the rotation of a coil in a magnetic field. The resulting flow of electricity peaks and falls as the coil turns, creating cycles of energy. The frequency of the current is basically measured by the number of times a coil makes a complete cycle within a unit of time. We were particularly interested in whether our five dimensional scores would cycle in a similar way.

In order to get a clear picture of what this would mean, think about a stadium full of sports fans doing "the wave." Each fan contributes to the pattern by quickly rising and sitting down again in sequence. Now, let's say we assign every fan a number reflecting his or her position in that simulated wave—1 for being seated, 2 for beginning to get up, 3 for being halfway up, 4 for straightening into position, and 5 for standing. We'd get a sequence that looked like this: 1, 2, 3, 4, 5, 4, 3, 2, 1, 2, 3, 4, 5, and so on.

Our idea was that the scores on our five variables might take the form of waves reflected by scores that rose and fell regularly in this way. We were picturing them as successive crests of emotional energy propelled through a dialogue. By treating each of the variables separately for both speakers, we could determine the frequency of changes each underwent (how long it took to go through one cycle), their respective cyclical power in the dialogue (how often the cycle was repeated over a session), and the influence the speakers might have on each other in this domain.

Harmonic Influence

We had already determined by our cross-correlation study that speakers influence each other temporally—sometimes with a time lag. Certain changes in our variables had consistent time-locked antecedents or after effects. But influence can also be reflected harmonically—that is, by a similarity of frequency between one dimension's cycles and another. For example, if the negative tone of a client's images peaks and falls at the same or related frequency as the therapist's introduction of new themes, the two items would be considered *coherent.* If the peaks and falls in these two cycles occur simultaneously, the two items would be considered *in phase.*

Part of what we wanted to know, by way of our five variables, was whether our speaking pairs were in harmony with each other. We wanted to know whether certain dimensions were more likely to cycle together than others, and whether there was any relationship between effects that move linearly through time as opposed to those that cycle through time.

SPECTRAL ANALYSIS

The mathematical method we used to explore these various issues is called *spectral analysis,* which establishes an object called the *power spectral density* (PSD) function. This function characterizes the strength of a signal's frequency, such as an electrical signal, along with the signal's overall power—that is the amount of energy delivered to the process during a given unit of time.

Think about the electrical current that comes into our homes. This is energy in motion. When we flip a switch, billions of electrons move like water into an appliance or device, which converts the energy into some form of work: heat, light, sound, and so forth. The power of an appliance is the rate at which it produces work within a given time period. An iron, for example, uses more power at higher settings than it does at lower ones. It consumes more energy to produce a greater degree of heat. The more power a machine has, the more energy it consumes, and the more work it can do.

These are the concepts we were using to measure our five dimensions. We regarded their changing scores as fluctuations of emotional current running through a session. The more powerful the dialogue, the more the current would surge, supplying energy for the work being done. Thus, in the logic of the model, the more an item varied from low-to-high and high-to-low scores, the greater the emotional energy expended in the exchange. The most powerful sessions would be characterized by variables that cycled over a wide variety of scores.

We ran the analysis separately for each dimension of a speaker's productions, but we also developed a total power score for all five dimensions in a session.

RESULTS OF THE ANALYSIS

Cyclical Power

Our first result was consistent across the board. There were no significant cycles in any of the sessions for any of the five dimensions. Emotional current, as we were defining it, was clearly DC in our six core consultations. It moved in one direction and was likely to stay where it was a short time ago. We did find some score patterns that were occasionally repeated, but they didn't contribute much to the power of the session. For example, a speaker might generate throughout the dialogue a stream of midrange scores—such as 3, 4, 2, 4, 3, 2, 4—but every so often, the stream would be punctuated by a surge or falloff of 1 or 5. These occasional rises and falls are cycles, but they didn't happen often enough to be significant in comparison to the DC trend.

Total Dimensional Power

Although the sequence of numbers did not cycle, technically speaking, the scores did—in the sense that a regular movement from high to low and back again indicated cyclical shifting from narrative image to intellectualization and from positive to negative tone. Thus, our model defined as a powerful sequence of scores the ones in which regular variation occurred.

One can see the logic of this by thinking about surges of power in an electrical current to supply energy to an appliance. Let's say we're using an iron to touch up some polyester shirts. We set the iron at level 3 for "permanent press" and leave it there until we're done. The current, in this case, is completely stable. The iron is always consuming the same amount of energy to deliver the same amount of heat.

But what if we keep changing the settings to accommodate different articles of clothing? We might iron a shirt at setting 3, go to level 6 for a pair of denim jeans, shift to 1 for a half-slip, then move up to 7 for a linen handkerchief. This isn't a particularly logical way to iron, but it should be clear that the varied pattern would require more energy.

In the same way, if a speaker's scores for positive imagery are consistently 2 or 3, the same amount of energy is required to generate the vehicles throughout the session. But if the scores on tone vary a great deal—now 2, now 5, now 1, now 3—one can assume the use of more emotional energy and a greater degree of communicative work. The key idea is that increased variety in scores calls for increased energy to support it.

What we found was this: For both clients and therapists, the most powerful dimensions were the amount of narration, negatively toned images, and newness of themes. Much less powerful were positively toned images and continuity.

This struck us as a significant result. The most powerful items—the ones that varied most as a session unfolded—were those most critical to shifts to and away from encoded communication (narrative images that have both conscious and unconscious meaning). It seemed reasonable to hypothesize that sessions in which the client keeps shifting from strong, emotionally charged narratives to intellectualizations and back again are more powerful than those that do not.

Total Systemic Power

When the power scores on all five dimensions were added together, the highest for *both* clients and therapists were achieved in the sessions conducted by Therapists B, E, and A. These were the three sessions that had obtained the highest cross-correlation scores, concentrated in the same areas of narrative imagery and new themes (for Therapists B and E) and negative images (for Therapist A). Thus, the finding integrated very nicely with our present understanding of deep influence, and it supported some of the hypotheses we'd developed about the relationship between influence and speaker stability.

Of all ten interviews, the sessions with Therapists B and E had been the most stable with regard to speaker alternation. That is, both therapists and clients in these two interviews had shown the strongest tendencies to hold the speaker role when they had it. Structural stability, influence, and power were all coming together as part of the same package.

Therapist A again was a bit of an anomaly, but the results we were getting were now becoming easier to understand. Although her session was the third-highest in influence, it held this position largely because the bottom three sessions had registered virtually no effects at all. Of our ten consultations, hers was one of the least stable. She and the client both tended to make short comments and to interrupt each other.

At this point, however, we saw a familiar pattern. The most powerful dimensions in her session, with respect to the PSD score, were the same dimensions that had shown a moderate degree of influence in our cross-correlation study—positive and negative tone, with an emphasis on negative. An unstable dialogue can apparently support an emotionally powerful exchange, but perhaps only in a limited area of communicative effects.

To summarize, spectral analysis had supported a number of the ideas we'd developed about temporal influence. The items that had the most impact on the other person's selection of communicative vehicles in a given session were the same ones that the new model defined as powerful. Newness of themes, amount of narration, and negatively toned images were emerging as important factors in human communication. We were beginning to have a sensible map of a small segment of deeper nature.

Of course, it should be stated that our results offered no particular basis on which to argue that powerful and influential exchanges are optimally or necessarily therapeutic. Our own suspicion was that a session without stability, influence, and power—as we'd defined and measured them—might enable a client to build better defenses against unconscious emotional responses. A stable session that displayed power and influence in these terms presumably allowed for more encoded communication and actual change. But these were simply conjectures on our part.

Communicative Coherence

Our final measure involved the question of coherence—the extent to which the scores of individual speakers might rise or fall in concert. Given our cross-correlational findings in the temporal domain, the results of this analysis were a surprise.

No coherency existed in any of our core consultations. Being related in the time domain was no guarantee of relatedness in the frequency domain. This result struck us as curious, but we put it aside for the time being.

A BIT OF SERENDIPITY

As we had before, we used our three couples' dialogues as a kind of control study, subjecting them to the same kind of analysis we'd conducted on the consultations. The scores in these nontherapeutic exchanges behaved much like the ones we'd already run. There were no cycles, but the narrative image and new theme scores tended to move from highs to lows and back again.

The fact that this general back-and-forth movement seemed consistent throughout the dialogues appeared to be an important finding. We wondered, given the parallels with deep influence, if the phenomenon was a function of dialogue itself. Would it occur in an emotionally charged individual speech or sermon? In order to answer this question, we asked three of the people in our couples study to sit alone with a tape recorder for forty-five minutes and to produce a monologue about his or her emotional concerns.

We scored the dimensions of these monologues in the same way we'd scored the dialogues and ran the numbers. What we found was completely unexpected. All three monologues showed significant cyclical activity on the PSD function—generally with three peaks at varying frequencies of repetition. Now we had a mystery on our hands. Perhaps communicative cycling was a natural propensity. Far from facilitating it, perhaps dialogue inhibited it.

In an attempt to get at this issue, we ran the scores from the three extra sessions we had for our cross-correlation study. These sessions, remember, were conducted by Therapist B and one of his colleagues. We had asked for these sessions because we were trying, at the time, to confirm the exceptionally strong results we'd had with Therapist B in the matter of temporal influence. Like Therapist B, the other therapist implicitly encouraged his client to talk for long periods of time, intervened at length, and interpreted some of the client's narrative images. All three additional sessions did, in fact, realize high cross-correlational scores.

Now we found that these sessions also generated significant cyclical activity—particularly for the clients, but also for the therapists. As with the power scores, the cycles occurred largely in the areas of narrative imagery, new themes, and negative tone. It struck us that these results were, in part, due to therapeutic style. These therapists gave their clients enough time to establish monologues, thus generating what does appear to be natural cyclical repetition.

On the strength of these results, we began to think about the phenomenon of regular variation in new ways. If communicative cycling is a natural propensity, then it can be understood in terms of other cyclical systems in nature, whose regularity tends to conserve energy and allows it to be replenished. Variation would include not only an increasing repertoire of expressive states, but returns to prior states, serving both creativity and reliable communication.

We began to wonder whether a dialogue could become too varied, much as an electric circuit can become overloaded by too many appliances. We decided to look at this issue more closely by exploring questions of emotional energy in thermodynamic terms.

CHAPTER TWELVE

Smashing Pumpkins

If the mind is, therefore, an energy state in which the increase of information is generating an increase in entropy in the surrounding system, then, for all practical purposes, the mind has to be looked upon as a very real event in the physical system.

—William Irwin Thompson, *Passages About Earth*

The title of this chapter refers to a comedy routine on David Letterman's *Late Show,* in which the host drops pumpkins or other breakables from a high-story window and watches with glee as they smash to bits on the street below. The humor of the piece comes, in part, from its subversive nature. But the routine is also funny because its outcome is never in doubt. Toss a pumpkin out the window, and it becomes Humpty Dumpty. Neither king's men nor king's horses can put the pieces together again.

Smashing pumpkins is a pretty decent illustration of the *Second Law of Thermodynamics,* which informed our next series of projects. According to that law, all isolated systems become increasingly disordered with time. Put differently, they lose some of their ability to go on working. The ongoing movement—from structure to disorder—is what physicists call *entropy*.

Entropy is apparent in every domain of life. Neglect the filing, and an office is soon overrun with documents that no one can find. Stop cleaning a house, and pretzel crumbs accumulate between couch pillows and the kitchen table buckles under unsolicited catalogs. Left to its own devices, an isolated system's entropy just keeps on growing. This is because the possibilities for disorder always outnumber the possibilities for order.

In fact, as Letterman's routine so nicely demonstrates, increasing the entropy of a system is easy. All we need do is drop a pumpkin from the nearest window. Decreasing entropy is another matter. It takes work to reconstitute a system. Without it, entropy is irreversible. The system will stay where it is or get worse.

In the nineteenth century, when the concept of entropy was first developed, it didn't refer to disorder as such. It referred to questions of heat and temperature. All physical systems take in some kind of energy from the environment and use it as fuel. As the process occurs, energy becomes expressed as motion and change, and the temperature of the system rises. The heat it gives off is a form of "free" energy. When systems of two different temperatures come into contact, heat flows out of the hotter substance into the cooler one until a uniform temperature is reached. Cream, in this respect, doesn't cool a cup of coffee. The hot coffee is literally transferring some of its energy to the cream.

What's actually happening here is a meeting of molecules. Hot coffee is hot because the individual molecules that make up its properties are in relatively vigorous motion. These vigorous molecules collide with the less active molecules of the cream, transferring some of their energy and also losing some of their momentum. The cream's molecules, meanwhile, reel from the collisions and move a little faster.

The result is like some kind of molecular bumper car ride. Until equilibrium is reached, the interaction is driven by differences in kinetic energy. The Industrial Revolution occurred when scientists discovered how to take advantage of this energy transfer. An engine, for example, converts the kinetic energy of exploding gases into mechanical work.

In all systems, however, some of the heat generated by colliding molecules is simply discharged and lost to the environment. Over time, this degrades the system's parts, decreasing an engine's capacity to do the same amount of work with the same amount of fuel. Entropy was originally intended as a measurement of such decrease—the diminishing capacity of a system to convert energy into work. Heat, energy, and work are all aspects of the same phenomenon.

EXPANDING THE CONCEPT

Ludwig Boltzmann, whose ideas underlie the project we were about to undertake, took the definition of entropy beyond questions of heat and energy. He did this by considering the impact of the constant combination and recombination of molecules that occurs during heat transfer on the system's original state of organization. The capacity of a system to do work, he said, is equivalent to the state of its organization.

To see this clearly, think about a big storage box full of crayons, organized according to color. What if we put a lid on the box and give it a shake? Obviously, the crayons will move out of place and collide with each other. Now the box is in a less ordered state. Shaking the box again will increase the disorder even further. Within three or four shakes, the arrangement will be irretrievable. Without work from an outside agency, the crayons won't get back to their original state of organization. This is pretty much what happens to a dropped pumpkin. The impact of the fall rearranges the pumpkin's molecules to the extent that its properties are no longer well organized forms of energy.

Once the molecules of a system are completely disorganized, they're pretty much like the disordered box of crayons. Any combination is just as likely as any other, and the overwhelming majority of these combinations give rise to a disordered manifest state. One might say, in this respect, that nature favors entropy. For example, large-scale weather systems become random as they make contact with more complex local systems.

Moreover, the very attempt to decrease entropy—say, by physically putting our crayons back in order—actually increases disorder in the world. Why? Because we, too, are energy machines. As we convert the organized energy of our food into work, we actually are increasing our own entropy.

This, in fact, is the situation described by the First and Second Laws of Thermodynamics; the first is called *the conservation of energy*. It states that energy can neither be created nor destroyed. It can only be converted into different forms. Thus, every attempt to decrease entropy in one part of a system inevitably increases it in another.

ENTROPY AND COSMOLOGY

The first two laws of thermodynamics may sound pessimistic—rather like saying that the only sure thing in life is death. But this is because science looks at such phenomena from the perspective of finite time. Sometimes it takes a mythology, whose temporal perspective is different, to give us another perspective on the same information. In the Hindu cosmology, for example, the varied forms in the universe are roles played by an infinitely changing, but eternally existent god—much like energy. The god's creative momentum is cyclical. Its "frequency" is the breathing of Brahma.

As Brahma breathes out, the cosmos expands into an infinitely complex variation of forms; as he breathes in, creation loses its impetus, and things fall apart—until the next breath cycle. A thousand such cycles constitute one day in the life of Brahma.

The thermodynamic view gives us a more concentrated view of the same cosmological phenomena. But, like the ancient texts, it also tells us how nature accomplishes creative variation. Entropy is the price we pay for the conversion of energy into new forms. Like breathing in and breathing out, like sleep and dreaming, the creative process in nature is necessarily cyclical. Cycles preserve the structural integrity of a system, but they also keep alive the potential for novelty and change.

This is precisely what we were finding in our measurements of communication—alternating stages of repetition and variation. Like the great wheel of tides and seasons, the movement of a dialogue is not a purely linear affair. It's more like a spiral, cycling in a way that is stable and predictable, but retaining the energic potential for influence and chance combinations of elements.

INFORMATION THEORY

In the late 1940s, an electrical engineer named Claude E. Shannon recognized that the results Boltzmann had arrived at with entropy and disorder could be applied to information systems as well. That is, he showed that entropy could be used to measure the complexity of a pattern of communication. It is this understanding of entropy that gave us a way to measure the information content of our own data.

It should be recognized that information is not synonymous with meaning. Information denotes the signals, electrical impulses, or channels used to transmit a message from a source to a receiver. The theory that Shannon developed can be used to measure the complexity of any informational code, including nerve impulses and brain signals as well as the transmissions generated by communication devices, such as the telephone and the radio.

One of the ways a communications engineer can ensure that a message be transmitted accurately is to keep its entropy low. This keeps extraneous signals out of the message. Another way is to use redundancy—to send the same message more than once.

What we hoped to do was to consider our five-dimensional data in terms of these particular concepts. The fact that we had found cycling of one sort or another in all our speakers' vehicles of expression suggested a form of natural redundancy built into our communicative patterns. But we had also seen nonrepetitive variation, which increased the complexity of the messages, their capacity for disorder, and the possibility of noise (unwanted information). It seemed to us that by measuring the complexity, in Shannon's sense, of our score trajectories, we would get a sense of the "emotional entropy" in our therapy sessions.

Our hypothesis was that a certain degree of entropy is warranted in a therapeutic dialogue—at least for the client. A therapy session, after all, is a communication system established for a particular purpose. A client normally seeks out a therapist to effect a change in an accustomed psychological pattern. The therapeutic relationship would theoretically open this pattern to the possibility of variation, portending a new form of organization. We speculated that the process could be observed in the increasing complexity of communicative vehicles in a dialogue. We already had strong indications that changes in these vehicles were related to the exchange of emotional energy between speakers.

As the systems theorists tell us, increasing complexity leads to a condition of random disorder only when the process is unchecked and the structure breaks down. Disequilibrium may, in fact, be necessary before a new structure can evolve. For example, there is a great difference between tossing a pumpkin out a window and using it to make a Thanksgiving pie. Both entail smashing the pumpkin and increasing its entropy, but making a pie claims the pumpkin for another kind of system. It also requires work to create the new organizational entity.

This is what we were looking for in our data—evidence of the emotional work that increases creative variation, as reflected by a speaker's vehicles of expression. We wanted to know what kind of therapeutic communications support this kind of work, and, in particular, whether the combined client/therapist system ever settles into an equilibrium. That is, does the transfer of emotional energy between therapist and client eventually reach a point of uniformity, so that change is no longer possible?

THE RESEARCH STRATEGY

In order to get at these questions, we first looked at the information in our dialogues simply as communication signals. We treated the speakers separately, because we assumed that therapists and clients would communicate with different degrees of informational complexity. In this way, we could determine how each participant contributed to the overall complexity of the client/therapist communication system. By complexity, we meant the sequence of changes the dimension scores underwent over the course of a session.

To get a sense of what this means, let's consider two sets of data. Client A has a series of narrative image scores that look like this: 1211222221. Client B's scores look like this: 11112222. Given that both sets have four 1's and four 2's, they might be considered equally complex. But we were defining complexity in terms of progression over time. On this basis, the first set is more complex than the second. It takes more rules to define it, and it would require more work to make it structurally regular—1212121212.

We would therefore conclude that the entropy of Client A's

sequence is growing more quickly than Client B's. It should be clear from this illustration that Shannon's ideas were entirely relevant to the domain of narrative expression.

Consolidating the Five Dimensions

Because of the small number of scores involved, it seemed unproductive to track the complexity of each dimension as a separate entity. We decided, instead, to treat our five dimensions—narrative imagery, new themes, positive tone, negative tone, and continuity—as a single item, a vector with five dimensions. Picture a five-outlet electrical adapter, into which five different devices are plugged: a TV, a radio, a tape deck, a CD player, and a clock, for example. None, any, or all five may be "on" at any given time. Noting which are "on" from second to second corresponds to the movement of the vector through time. The five numbers would represent the status of the speakers' communicative vehicle at any given time.

We called this five-dimensional unit an *Information Particle* (IP), because we considered it the elementary constituent of the phenomena we were studying. We used software that would note changes in any of these five numbers every eight seconds of dialogue. Suppose, for example, that a client obtained the following set of scores for the first eight seconds of a dialogue: Narrative Imagery 2, New Themes 2, Positive Tone 1, Negative Tone 0, Continuity 3. The sequence would look like this: 22103. Let's say the scores for the next three eight-second increments are:

Seconds 9–16: 22102
Seconds 17–24: 22103
Seconds 25–32: 22103

It should be clear that this sequence of scores is repetitive. Only one new state occurs within the thirty-two seconds tracked. Thus, the sequence has a low entropy value. This is the sort of progression that might occur at a point in a dialogue where the client is simply giving the therapist basic information or speculating about why she fell ill—intellectualizing.

Now, let's say that the same client obtained the following scores in the same four increments of dialogue with another therapist:

Seconds 1–8: 22103
Seconds 9–16: 32112
Seconds 17–24: 42013
Seconds 25–32: 42213

Here, each increment in the sequence involves a new dimensional state. The scores indicate a shift from intellectualizing to narrating, with a new and strongly negative thematic thrust. The progression varies more, so its entropy is higher. Because speech lengths differed from session to session, we programmed the computer to sustain the last scores recorded if a speaker was silent. That is, we took the silent speaker to be in his or her last observed state until the silence ended.

In addition to tracking sequential complexity, we intended to use the data points to generate a curve that would reflect the path of a speaker's communicative entropy throughout the session. A relatively flat curve would indicate little change in complexity during the hour, whereas a steeply ascending curve would reflect rapid increments in variety of expression.

It should be emphasized again that the variables we were tracking reflect the speakers' manner of expression—the *way* in which they communicated. Each combination of scores was a configuration of signals designed to transmit a message. A key question was whether the Second Law of Thermodynamics, known to apply to all observed material systems, also applies to communicative, immaterial systems.

Applying Shannon's Theory

As I said earlier, in information theory, entropy is a way of measuring the variety or complexity in a pattern of information. Beginning with Boltzmann's ideas on material systems, Shannon developed a mathematical formula to calculate the entropy of a particular message. The greater the complexity of the message, the higher its numerical value. If a message has no complexity, its entropy is 0.

For example, the one-word message "Yes" has no informational complexity. The pattern of information that transmits it lends itself to no other possible message. "Yes Yes Yes" would be redundant. The complexity value of the information would remain 0.

We analyzed our sequence of dimensional states by applying Shannon's formula to them. The entropy value at any moment in the session depended on how often each combination of scores already noted had reappeared in the data.

Let's go back to our sample series from the first four increments of a dialogue: 22103, 22102, 22103, 22103. Given the performance so far, the probability that our next eight seconds of dialogue will be 22103 is very high. The same type of communicative vehicle is being invoked again and again. As a speaker narrates, the contents of the material might change, but the speaker's manner of expression is consistent. Entropy here would be minimal. The maximum state of entropy would occur if each configuration of the five dimensions occurred only once or equally often in an entire session. This would make the reappearance of any of them equally likely.

In general, information and entropy increase simultaneously. It thus seems reasonable to expect a client to register more complexity than a therapist. Beyond a certain level, however, increasing entropy would theoretically decrease a Client/Therapist system's capacity to make use of the information available. We figured that a middle range of entropy values would indicate a dialogue rich in usable information.

Calculating the Odds

It should be noted that we were tracing entropy cumulatively. Every new state increased the options we were calculating. A sequence like the one just described has but two possibilities—22103 and 22102—and is relatively easy to configure. But every new state increases the options and the complexity of the system. Given the fact that we had scored Narrative Images from 0 to 5, New Themes from 0 to 4, Positive and Negative Tone from 0 to 3, and Continuity from 0 to 4, there were 2,400 possible combinations in all (6 x 5 x 4 x 4 x 5 = 2,400).

The maximum value that entropy can achieve in a study of 2,400 possible states is 7.78. Midrange, therefore, is somewhere near 4. Values lower than 4 would imply the repetitive use of a few communicative vehicles; values above 4 would suggest the use of so many different modes of expression as to subvert the dialogue's structure.

CHAPTER THIRTEEN

The Law of Emotional Entropy

No law can be sacred to me but that of my nature.

—RALPH WALDO EMERSON

AS WE HAD in our other trials, we selected our first scores for analysis at random—those of Client 1 in her session with Therapist A. This was the consultation in which an empathic, talkative female analyst had interpreted the client's problems in terms of an exterior "caretaking" self and a neglected child within.

We ran our five-vector series for the client, then plotted the changes on a graph. The resulting curve, which showed her increasing selection of new states across the session, was unexpectedly smooth. It was so smooth that it suggested an *underlying determinism.* As it turned out, the number of communicative options in play had grown so smoothly that it could be calculated as a *logarithmic function of time.*

When something increases logarithmically, the rate of its growth can be determined in a precise way by the amount of something else. For example, the time required for a bacterial culture to reach a given size varies as the logarithm of the size of the original colony. (One might even consider the logarithmic nature of one version of Murphy's Law, which states that work always expands to fill the time allotted to do it.)

In our data, the entropy of the client's communications formed a near-ideal logarithmic curve when we plotted its growth against the number of seconds elapsed in the session. The scores for the therapist gave us the same kind of results. The real data was a near-perfect fit to the ideal curve. We quickly ran the other ten trajectories and found that the log curve existed for all twelve, clients and therapists alike.

There were strong individual differences, of course. The greater the complexity of a speaker's communications, the more steeply the curve would trend upward. Each curve had its own distinct character. Nonetheless, all twelve curves accorded perfectly with the Second Law of Thermodynamics. In each instance, entropy had grown with time. Indeed, the law did more than hold; it held in a specifically logarithmic way. The finding argues for a universal mental structure that governs the selection of communicative vehicles when we express ourselves in a time-limited dialogue.

By converting the pattern into an equation, we could predict the individual differences that occurred within the constraints of the law. Our predictions held up for the psychotherapy trajectories, our couples' trajectories, and even the ones generated by monologues. People apparently vary their manner of expression in a way that is both universal and lawful. As time elapses, new states of communication enter the system at an ever-slowing rate.

Every pattern in our study was the same. During the first few minutes of the time period, there was a burst of expressive activity. The speaker used many varying configurations of the five-vector vehicle within a short period of time. Then the rate at which new states were invoked slowed down, and the speaker returned more often to states used before. No speaker, however, settled into a pattern comprising only prior states. All parties were trying out new configurations of the five dimensions to the end of the period. The initial burst served to set a basis for the communication, and later additions enlarged it.

SOME IMPLICATIONS OF THESE RESULTS

The strong fit between ideal curves and real values for every one of our trajectories struck us as highly significant. In a consultation session, the speakers have just met, and they are exchanging a great deal of new information very quickly. So we expected to find that therapists and clients would vary their manner of expression in the initial minutes of an interview. We did not anticipate, however, the strong adherence—for all speakers—to a logarithmic pattern of complexity.

Despite the differences in the speakers' roles, their ways of communicating, the range of their surface comments, the variability of their subject matter, and the multiple ways in which they influenced each other, entropy always followed the same pattern. Indeed, the rhythmic alternation of new and prior states throughout the time period suggests a fundamental process of the mind—one that must link up to basic properties of the brain as well.

It is perhaps significant that recent studies on Alzheimer's disease have shown that people who eventually develop the condition are marked in youth by a rather flat style of communicating, lacking in variation and tone. Alzheimer's has since been linked with abnormal brain glucose metabolism. The human mind has apparently evolved with a natural propensity toward complexity and novelty, whose impoverishment can tell us something about distinct neurological conditions.

In order to verify the results we were getting, we ran a statistical study of two types of random data. The entropy of the random data did not grow logarithmically, so the log growth appears even more specific to the human system than we may have first guessed. The way we express ourselves—the kinds of stories we tell, the images we use, the variety of themes, their tone and continuity—is fundamental, more fundamental than what we actually say.

Clearly, the law we were seeing—we called it *the law of emotional entropy*—belongs to nature. There is no conceivable way to be aware of or to control a logarithmic increase of expressive states. Like other isolated systems, the Client/Therapist system is closed—for its duration—to outside influence. Thus, an increase in entropy is natural and irreversible. As time elapses, the relation of energy to information becomes more complex, more random, and less efficient.

Although it might be objected that a speaker does periodically return to configurations of the five dimensions already used, this reentry into prior states does not mean the process is reversible. When a prior state is repeated, it is repeated after other things have entered the system. Hence, there is no restoration of an original state of organization.

Ultimately, of course, the people creating a Client/Therapist system can depart from each other and replenish the energy of their individual systems, just as a laborer can go home, eat, sleep, and wake up the next day with a renewed capacity for work. It's even possible that, in a communicative system, the speaking parties enable replenishment for each other during the course of the exchange.

THE AVERAGE LEVEL OF EMOTIONAL ENTROPY

FINAL ENTROPY: THE CLIENT/THERAPIST SYSTEM

We were interested to discover that the cumulative entropy value for every Client/Therapist system was in the midrange, hovering around 4. (The possible extremes were 0 and 7.78.) Nature would appear to determine a consistent mix of informational redundancy (a repetitiveness that ensures stability and the ability to reconstruct the messages being sent) and creative variation (an exploitation of informational possibilities that increases instability and disorder). Given that consistency, we wanted to see how the individuals in each case contributed to the cumulative score.

Final Entropy: The Clients

There were only two clients represented in the cases we were using, but each had seen three different therapists. So we had six client trajectories to analyze. The differences were radical.

Of the six, the steepest curves—indicating the highest degree of complexity and variation—were obtained by Client 1 in her sessions with Therapists B and C (the relatively silent therapist and the therapist who had spoken with the client before the videotaped session began). This same client also produced one of the flattest slopes of the six—in her session with Therapist A, the very active female therapist. These striking differences in final entropy in a single client strongly suggested therapist dominance in the matter of expressive choice. We were particularly interested in the high complexity score produced with Therapist C. Our cross-correlational studies had shown this system to have been almost without effects in the matter of mutual influence.

Client 2 had much lower scores than the ones produced by Client 1. Her highest was for the session with Therapist F, whose approach had been somewhat intellectual. With Therapist D, who had been challenging and unpredictable, she produced the flattest slope of all six. Her final entropy score with Therapist E was in the low middle range, which also surprised us. Other models had shown this consultant to have elicited some powerful images, and the system had displayed more deep influence than others.

Final Entropy: The Therapists

Of the six consultants, Therapists A and D had the highest entropy scores. Therapist E was next. Therapists C and F were on the low side, and the lowest by far was Therapist B, whose long silences had been registered by the software program as repeats of the last configuration recorded.

Getting the Whole Picture

In order to better see what was going on with each speaking pair, we decided to get a ratio between the cumulative entropy values of the client and therapist in each session. The higher the ratio, of course, the greater the difference between one speaker's complexity values and the other's.

The greatest difference occurred between Therapist B and Client 1, followed by Therapist C and Client 1, and Therapist F and Client 2. The lowest ratio occurred between Therapist D, the unpredictable consultant, and Client 2—less than 1.0. In fact, as the ratio indicates, this therapist's cumulative entropy was higher than the client's. The ratios for the pairings with Therapists A and E were also on the low side.

As we worked with these scores, we realized that the clients who had the highest entropy values had invariably been seen by the therapists who had the lowest. The reverse was also true. The clients who had the flattest entropy curves had been seen by the therapists who had the steepest.

LOWEST ENTROPY VALUES		HIGHEST ENTROPY VALUES
Therapist B	⟷	*Client 1*
Client 2	⟷	*Therapist D*
Client 1	⟷	*Therapist A*
Therapist C	⟷	*Client 1*
Therapist F	⟷	*Client 2*
Client 2	⟷	*Therapist E*

This pairing was so consistent that we were forced to assume that complex and varied forms of expression from a therapist elicit a high degree of redundancy from the client and a low degree of novelty. One might even suggest that a highly varied and active therapist actually pushes the client toward repetition and away from innovation. Conversely, the redundant therapist, who intervenes less often and whose manner of expression does not shift from state to state, appears to create conditions more favorable to high levels of complexity in a client. This agrees with the view of systems theorists that systems tend to naturally balance extremes.

The results with Therapists C and E may be most striking in this regard. The system with Therapist E and Client 2 had displayed a fair number of cross-correlational effects in our Measure of Deep Influence, and we had credited them to the pressure he put on the client to tell stories. Yet, the high complexity of his interventions had driven down the client's inclination to venture far from familiar expressive territories. The system with Therapist C and Client 1, on the other hand, had shown very few effects in the cross-correlation study, yet the client achieved one of the highest levels of complexity on the entropy model.

There would appear to be an optimal level of total entropy for human systems in a charged dialogue. If one person contributes a great deal to this quota, the other may be obliged, at a deep psychobiological level, to maintain a stable communicative base by repeating the same expressive state more often. Conversely, a therapist who is more redundant or attentively silent may implicitly permit the client to be more inventive.

The Narrative Question

We began to wonder whether high levels of informational complexity are related to high levels of storytelling. Had the therapists with high entropy scores contributed more narratives to the dialogue than the therapists with lower scores?

When we checked averages for newness of themes and extent of narration, we found that Therapists D, A, and E (the consultants with the highest final entropy values) had indeed contributed the highest amounts of narrative imagery to the dialogue. They had also introduced more new themes than the other speakers. Somehow their use of stories and new material had reduced the variability of their clients' vehicles.

Free association, that great hallmark of psychoanalytic technique, appears to be anything but free on the level of deeper nature. It is remarkably lawful and influenced to a surprising degree by the complexity of the *therapist's* communications.

UNTAPPED POSSIBILITIES

Given the fact that our entropy values reflected three basic propensities in a speaker—the total number of new states used during a session, the tendency to return to favored states, and the type of states visited most often (stories or otherwise)—we looked more closely at them.

It may be recalled that the total number of configurations possible with our five-dimensional Information Particle was 2,400. Some of the combinations are admittedly unlikely, but, even so, the speakers in our study used remarkably few. The range was from 95 to 128 different combinations for the clients (they averaged 117 per session) and 52 to 116 for the therapists (they averaged 80 per session).

This represents, at most, about 5 percent of the states available. The clients, in general, displayed a wider range of state selections than the therapists, with the notable exception of Therapist D. In general, as indicated, the more states visited by a therapist, the fewer used by the client.

Interestingly, in each ten-minute epoch, all the therapists actually produced more informational variety than their clients. However, their cumulative entropy scores (over forty-five minutes) were generally lower because they were more inclined than their clients to cycle back into using old states. This was particularly true of Therapist A, whose manner of intervention was highly complex, but alternated with a consistent return to favored vehicles of expression.

We also discovered that most repeated state selections, for both clients and therapists, were intellectual and nonnarrative in character. The one exception was Client 1, whose favored states were predominantly narrative in the sessions with Therapists B and C. She and Therapist B showed the least inclination of all the speakers to cycle back to any favored states.

Even so, the average of the actual scores for each dimension made clear the nonnarrative character of much of these dialogues. Low values for narrative imagery, new themes, and positive or negative tone by definition indicate intellectualization and redundancy. We found them consistently lower than average in all our trajectories. The exception was the dimension of continuity, whose values were consistently higher than average. High continuity values, however, merely indicate a tendency to stay with the same themes over the course of a dialogue.

Thus, both parties to therapy, at least in these six sessions, favored nonnarrative communication and were disinclined to expend the emotional energy required to sustain encoded modes of expression. This was one more indication that only a small segment of the 2,400-item pool was being used at any given time.

If one were to picture our Information Particle literally as a material entity, it will have occupied only a very small region of an enormous five-dimensional space. And much of that region was a low-score, intellectualized space, as opposed to a high-score narrative space.

TRANSLATING OUR RESULTS INTO THERMODYNAMIC TERMS

The concentration of scores in combinations that reflect low levels of narrative expression can be accounted for thermodynamically. Unencoded modes of expression may be our natural, resting, low-energy state. Movement into unconsciously laden, encoded communication (stories) is a bit like climbing in a higher altitude, which expends more energy. We are likely to sustain the effort for only short periods of time.

In order to minimize the expenditure of energy, we make greater use of nonnarrative modes of expression. Moreover, given the fact that higher levels of entropy are associated with higher levels of narrative, excursions into encoded communication are probably change-associated and irreversible. The mind may avoid frequent entry into such a space as a way of precluding systemic overload and disintegration.

Perhaps both parties in a dialogue remain in nonnarrative space until some inner need or external pressure prompts them to expend emotional energy in narrative expression, momentarily destabilizing the communicative system. In a therapy session, this experience can either be disruptive or promote growth and change. As the energy needed to sustain encoded communicative vehicles gives out and anxiety mounts, there is a shift back to unencoded space, where energy can be renewed for the next venture into the encoded domain.

These ideas may help to account not only for the many intellectualized, nonnarrative forms of therapy that are currently in vogue (they may help clients to build stronger emotional defenses), but also for the techniques, both Eastern and Western, that utilize active imagination, storytelling, and fairy tales to generate unconscious emotional effects. The role of narratives and images in mental economy evidently extends beyond the realm of psychodynamics into the domain of nature itself.

ENTROPY IN MOTION

We did one last study to complete our investigation of emotional entropy in communication. We had been considering only final entropy—the accumulation (to a final value) of configurations as a dialogue progressed. But we also wanted to know the dynamics of the pattern from second to second.

We had already looked at this aspect of our five variables as separate items. We had done spectral analysis to determine the power of their flow across a session and their tendency to cycle. We had also done the cross-correlation study to locate any temporal relationships between one speaker's dimensions and the other's. Now we used a moving window to track the entropic movement of the five dimensions as a single entity.

The Temporal Domain

First, we looked for cross-correlations between the cumulative entropies of the therapist and those of the client. As one speaker's five-vector configuration changed in complexity, would the other's do so as well? As we had earlier, we opened a ten-minute window so we could measure effects five minutes before or after a given change of score.

As it turned out, none of our six cases showed any cross-correlated effects with respect to complexity. The items that had shown temporal influence when measured separately displayed no such influence in combination. Given the fact that entropy had accumulated so lawfully, this was a striking result. The degree to which entropy changed from one second to the next was absolutely stable in both parties. The extent of prior change anticipated subsequent change. Yet, the movement of entropy in the two trajectories had no discernible relationship in time.

The Frequency Domain

We turned next to spectral analysis, looking for cyclical activity in the movement of entropy across a session. As we had with the separate items, we found that both speaking parties tended to expend the greatest amount of energy in remaining with a fixed information pattern. Repetitive cycles existed, but they were not powerful.

On the other hand, when we measured coherency—to see whether the cycles of therapist and client (however weak) were in sync with each other—we got a surprise. For all cases, there was a very high coherency between client and therapist. The speakers were cycling through varied configurations of expression in the same sequence. They simply weren't doing it at the same time. This is not unlike freestyle dancing, where both partners share a basic rhythm, yet express themselves in a very individual way.

UNEXPECTED IRONIES

When we had cross-correlated our five items as separate entities, we had seen a broad range of temporal effects. Yet, once the items were combined into a five-vector configuration, the temporal cross-correlations disappeared. Conversely, the individual items had displayed no coherency whatsoever in the frequency domain. Yet, once they were treated collectively, the coherencies were powerful and consistent.

This was a quantumlike riddle. Like fundamental atomic entities that behave as particles individually and as waves collectively, the individual elements of our Information Particle behaved in one way as separate items, and in another way together. Nature seems to employ the same principles of organization and behavior across very different realms. Mind and matter appear to be two sides of the same, natural universe.

THE NEXT STEP

Our data thus far had given us reason to assume that psychic energy is a real entity, whose effects we were measuring in terms of communicative output. We had seen lawful results when we measured communicative entropy by way of equations developed specifically for complexity of information. We wanted to analyze our data further by using what is called the *caloric model,* which is designed to measure entropy by way of heat transfer and work.

In general, the two ways of measuring entropy are not applied to the same system. The equation used to measure the entropy of a physical system involves two values: the amount of reversible heat absorption divided by the value of the absolute temperature over all epochs of change. Information doesn't lend itself to quantification in those terms. But we were convinced, because of the results we had obtained so far, that we could develop ways to treat the properties of human communication in the fundamental terms of classical physics—energy, force, and work.

I spoke earlier about picturing our Information Particle as a material entity moving in and out of various locations in a five-dimensional space. Our intention was to translate the vehicles of human expression, as represented by our five-vector Information Particle, into a geometrical figure. This conversion would give us a way to develop measures that have been heretofore defined only in the physical realm.

Like the bicycler whose journey can be converted into numbers on a two-dimensional graph, each speaker's Information Particle had a trajectory. It traveled to various locations within its five-dimensional box. To define its exact location, we treated each configuration of the items as a fivefold address. Each of the five numbers corresponds to a position on one of five axes that run though the box, one for each dimension. By analyzing the travels of the IP from one address to another, we could identify and quantify properties of human communication that are otherwise impossible to analyze. This project occupied us until we were satisfied that the results would serve us for a fresh measurement of caloric entropy.

CHAPTER FOURTEEN

The Physics of the Mind

Science is nothing but perception.

—PLATO

PSYCHOANALYSIS HAS long told us that some of our behaviors are dictated by processes beyond our awareness. The hard evidence we had accumulated indicated that this was true—not only with respect to unconscious fears and intentions, but with respect to our manner of expression, an aspect of conversation we normally don't even question.

It would seem that surface observation and subjective experience have little relationship to some of the things that are actually going on when we communicate with each other. Consultations that had struck us as prosaic and slow-moving were revealed by mathematical models to have been deeply interactive—to the extent that the speakers' changing configurations of expression occurred at the exact same frequency. Conversely, interviews that had seemed vigorously active and emotionally engaging had displayed, under mathematical analysis, few signs of mutual communicative influence.

We found this discrepancy fascinating, because it calls our immediate experience of a dialogue into question. Most people who are interacting in a way that an observer would describe as "emotionally engaging" feel subjectively that they're expending energy. They may describe the interaction as hard work, or talk about how difficult it was to get a point across. We wondered if these subjective feelings would translate over into formal measures of work and force as they are normally understood in the physical sciences.

WORK AND FORCE

Galileo first analyzed motion in terms of the distance traveled by an object and the time it took to get there from its point of origin. Sir Isaac Newton built his physics on these foundations, defining the properties of force, mass, work, and motion. His laws eventually had to be generalized for objects traveling close to the speed of light, but his three laws of motion remain true for physical bodies that move much slower than that. Thus, we felt confident in turning to Newton at this point in our research.

We had so far been analyzing our data with mathematical models that determine with probability, not certainty. Our research had suggested, however, that the behaviors of human communication are consistent with the behaviors of all physical systems that use energy. Thus, the data from a closed communicative system—such as occurs in a therapy session—should be measurable in classical physical terms. We decided to treat emotional energy as an agent that acts on the communicative vehicle so that we could quantify its activity in terms of mechanics.

In Newtonian mechanics, the action of a force on matter is called *work*. Work occurs, however, only when the force succeeds in moving the object it acts on. Newton formulated his definition of work in terms of an equation: W = fd, or Work equals force times distance.

In order to apply this equation to our data, we needed to find comparable elements in the communicative vehicle. This is why we converted our Information Particle into a geometric figure. The IP was our communicative object. It represented the communicative vehicle whose behaviors we were attempting to analyze. As stated at the end of the last chapter, by turning the IP into a geometric entity, we could regard its five numbers as positions on five axes. As the configurations change, the IP travels between different addresses in its five-dimensional box. We used these changes in position for calculations of distance, motion, and velocity (speed plus direction).

BASIC CONSIDERATIONS

Force, in Newtonian mechanics, changes the motion of an object. It is external to the body it acts on. Thus, we were conceptualizing a force acting on the Information Particle. We were not attempting to measure the force *of* the Information Particle. Ultimately, we pictured a *communicative force field* that propels human expression from one state to another. For our first project, however, we simply conceptualized a force that acts on human expressive tendencies—that keeps a speaker's IP in motion.

Anything that can move an object through space by expending some of its inherent energy is performing work. Work results in a change of motion, producing kinetic energy. In a closed system, this changes into other forms. By finding comparable elements in our Information Particle, we were attempting to determine the nature of the energy and its transformations in the Client/Therapist system.

Given the already noted equation that work equals the sum

of force times distance over the action path, the amount of work accomplished by a system varies. It depends on how much force is applied to an object and how far the object moves. Think about holding a ten-pound barbell one foot off the floor, then moving it upward to a height of five feet. Obviously, the distance involved is four feet. This is the "d" in our equation.

But the work required to move the barbell through those four feet depends on how much the barbell weighs and whether it keeps moving in the same direction. This is the force, the "f" component of our equation. By plugging our values into our equation, W = fd, we get the following result: 10 pounds x 4 feet = a work equivalent of 40 foot–pounds.

We were using our geometric IP figure to determine values of this sort. By noting the distance the IP moved from one set of coordinates to another, we could determine the values for "d" in the W = fd equation.

The Mass of the IP

The "f" part of our equation was a more speculative issue. In Newton's famous equation, Force equals mass times acceleration (F = ma). But what is the mass of a communicative vehicle? Would it vary, depending on its construction? Some combinations of scores seemed "more weighty" than others, but subjective considerations of this sort do not necessarily translate into differences in mass. If anything, we'd found that complex scores changed rapidly, as though they were "light," and simple scores changed slowly, as though they were "heavy."

Ultimately, we reasoned that questions of mass were irrelevant. Our primary objective was to use the numbers as coordinates so we could track the movement of the IP from one address to another. The numbers in 110 Baker Street have no more "mass" than the numbers in 914 Rodeo Drive. Thus, we decided to keep things simple and assume a fixed universal mass of 1. If the models didn't work properly, we'd go back and reassess the situation.

The Acceleration of the IP

The *velocity* of a moving object has two aspects: speed and direction. We were determining the velocity of the IP by noting its location at each second of the dialogue and subtracting from it the value of its prior location. This gave us the distance it traveled between the two positions.

But motion doesn't necessarily occur at a fixed rate of speed. A car traveling down a highway may do so at different rates of speed, even though it continues to go in the same direction. *Acceleration* is the rate at which the velocity of a moving object changes. It's calculated by dividing the change in velocity by the time it takes for the change to occur. The greater the acceleration, the greater the force that's causing the object to move.

We'd traveled this territory before, when we were using spectral analysis to determine how much our individual sequences of dimension scores varied. The greater the variance, the more powerful the sequence. In the same way, larger distances between states could be conceptualized as more powerful, thus requiring more work. High acceleration would indicate that the IP had traveled these distances very quickly.

Let's go back to the example of varying the settings on an iron, which came up in Chapter Eleven. Changing the settings from high to low and back again over a period of time is equivalent to using highly varied configurations of expressive states. More power is involved in varying expressive states than in keeping them the same.

Now we were measuring the distance an IP travels from state to state, which is comparable to measuring the distance between one setting on an iron and the next one used. Moving an iron from level 1 (to iron a half-slip) to level 7 (to iron a cotton handkerchief) entails a rapid delivery of energy to the system—more so than moving it from levels 1 to 2.

Similarly, the greater the distance and acceleration of an IP between one state and another, the more work was required for the state change and the more energy the speaker used to support it.

THE BASIC PLAN

We had already taken three steps in the direction of this project. We had:

1. selected and quantified five dimensions of human communication (attributes of the communicative vehicle);
2. combined the five scores to create the Information Particle; and
3. converted these IP scores into geometrical coordinates.

Now we were proposing to:

4. locate the IP in the five-dimensional box and track its travels from second to second;
5. determine the distance traveled by the IP for each second of the session, thereby obtaining the *velocity* with which the IP moved within its five-dimensional box;
6. convert the velocity values to *acceleration* (the change in speed divided by the time it takes to produce that change);
7. place the acceleration values into Newton's Law of Acceleration, $F = ma$ (Force equals mass times acceleration), and assume a consistent mass of 1 for the IP;
8. use the force equation and knowledge of coordinate changes to compute the work done on the IP and its kinetic energy; and
9. assume the existence of a communicative force field that moves the IP of the client and therapist in the course of a session.

JUSTIFYING THE MODEL

Clearly, we are arguing that plotting the IP in five-dimensional space is a legitimate analog of, for example, plotting the position of an aircraft in the nine-dimensional state vector used in navigation and guidance. An aircraft's position in three-dimensional space is calculated by determining three addresses each for position, velocity, and acceleration—nine dimensions in all. We intended to use our five dimensions in a similar way. The actual state of communication took place in the real world of three-dimensional space, but we were going to measure it in a five-dimensional communicative space.

We want to emphasize that the parallels we were seeing between physical objects and communicative vehicles are just that—parallels. They represent shared properties in different realms of nature. A physicist would speak of physical work—ultimately a change of state. We were talking about communicative work, which also results in a change of state. Shannon had already determined a precedent for this kind of enterprise, when he applied Boltzmann's equations for material gases to channels of communication. He essentially lifted the physical concept of molecular organization to the domain of information.

It should be recognized that physicists no more see the smallest particles of nature than we see an Information Particle. The state vector they use to infer the properties of subatomic particles is not unlike our five-dimensional box. Values for velocity and acceleration are obtained by indirectly observing and quantifying changes in position. The forces of nature that we take for granted—gravity, electromagnetism, and so forth—are postulates derived from the observation of material phenomena. We were attempting to understand changes in communicative phenomena in terms of another such force of nature. Such a force would be psychic, not physical. But it would suggest that matter, energy, and mind are equivalent, regulated by common laws. This, ultimately, is our cosmic circle.

THE FORCE FIELD

Although we had found that monologues had many properties in common with our dialogues, we decided to concentrate in this project on the six consultation sessions. The Information Particle represents the state of the two speakers with respect to five dimensions of their communication. Because these variables had already supported meaningful results, we regarded them as a small slice of a larger entity, but we had no illusions that we were measuring the total state of a therapy participant.

We were simply attempting to identify the force that gives energy to the speakers' Information Particles, moving them from position to position within the five-dimensional space. As an initial hypothesis, we reasoned that this energy must be generated by the interaction of the two parties—the communication system. Indeed, it made no sense to think of the pair as separate entities with isolated inner propensities when everything we knew about them arose in the context of their interaction. We therefore assumed a communicative force field that is jointly created by the client and therapist.

A Question of Measurement

In Newtonian physics, acceleration is expressed by measuring two changes in velocity. Because of the way our scoring method produced data, the application of Newton's equation could not result in a negative score for the amount of work performed on the Information Particle. At any given time, the amount of work would be defined as either unchanging or increasing. Negative work does exist in physical nature and is associated with loss of energy, as occurs when an object which slides across a floor loses heat to friction.

Although our scores are made every 8 seconds, force and work are measured on a second-to-second basis. So there are always scores that are carried forward and repeat, which result in a score of 0 for distance. Change, in this respect, is always a positive score. It would be impossible, however, to measure our five variables literally second by second. One would be scoring something like two words at a time. This limitation may be analogous to a limitation found in atomic physics: Microscopy cannot resolve details finer than a certain size without compromising the reliability of the measurement.

After we obtained our initial results, we did apply several mathematical techniques to determine the extent to which our scoring method was corrupting our findings. Interestingly, we found that only about 3 percent of the positions available to the IP could theoretically produce negative work. We also corrupted the data with several types of noise, wildly altering the scores so that negative work would easily appear. We still got little indication of negative work.

Frankly, we suspect that we are dealing here with a philosophical truism rather than a measurement problem. Communication always moves forward, never backward. One cannot unsay what has been said. Perhaps negative work is not a viable factor in a communicative force field.

CHAPTER FIFTEEN

Working Hard or Hardly Working?

Without knowing the force of words,
it is impossible to know men.

—CONFUCIUS

ONCE AGAIN, WE chose our first run of scores at random. We used the IP configurations of Therapist E—the consultant who had elicited some powerful stories from Client 2 about her suicide attempt. Given this system's moderately strong cross-correlational effects on individual dimension trajectories, we had been surprised to find that the complexity of the client's communications in this session had been rather low. Now we were plotting the therapist's cumulative work on expressive states—the work mediated by the communicative force field on the therapists' Information Particle—against time elapsed in the session.

Even to the uninitiated eye, the resulting curve was unmistakably linear. Once again, the real data perfectly matched an ideal curve. As it turned out, all twelve work curves were like this—as were the ones from our other sessions, dialogues, and monologues. The curve we saw was actually a straight line: Work and time had consistently accumulated in equal increments across the communication span.

In other words, the force that moves an Information Particle from state to state across a conversation is steady and unvarying. All that changes from session to session is the rate at which the work is done—how much work is effected per unit of time—and the level to which it ultimately builds. This individual characteristic is reflected in the slope of the curve—how steeply it ascends.

WHAT DO THE RESULTS MEAN?

Although speakers may feel that some emotionally charged encounters require more "work" than others, the actual work, as measured by changes in communicative states, does not appear to vary at all. It is evenly distributed throughout a dialogue. This suggests that a constant supply of work (and energy) is necessary for any kind of expressive variation. The discrepancy between subjective experience and formal measurement does not indicate that our feelings are mistaken, however. The subjective experience of emotional work appears to be the end product of another, deeper psychic process, with a different mode of stable expression.

Individual Work Rates

To get a better idea of what this result meant, we decided to look at the slopes of these curves and to consider the clinical implications of individual differences. As with our entropy curves, the degree of a slope indicated the specific rate at which work had accumulated for a particular speaker. The more rapid the rise of the line, the greater amount of work done to the speaker's Information Particle.

The first observation we made was that total work scores

were relatively close for both parties. With final entropy scores, high complexity for the therapist appeared to ensure low complexity for the client and vice versa. In the case of accumulated work, high-scoring therapists were invariably paired with high-scoring clients, and low-scoring therapists were paired with low-scoring clients. The range of accumulated work values extended from 6:5 (Client 2 with Therapist D) to 3:1 (Client 1 with Therapist B).

This finding encouraged us in our theory that client and therapist together create a communicative force field that affects their individual expressive states in comparable fashion. On the other hand, in all cases, the total accumulation of work was somewhat higher for the client than for the therapist. This finding is consistent with the fact that clients tended to use more new expressive states than their therapists. Any state change requires work.

To look at these results from a different angle, the slope of a work curve represents the power of the force field that determined the rate at which work is performed. Power is associated here with the moment-to-moment force acting on the five-dimensional Information Particle to move it from state to state. Given the definition of the slope as a fixed parameter, we can say that the instantaneous delivery of power to a Client/Therapist system was uniform over the course of each session. The amount of power actually delivered, however, differed from session to session. This difference was reflected in the size of the slope.

Unlike the results we had with accumulated entropy (in which high levels of complexity were generally associated with clients in the most stable systems), our accumulated work measurements showed that high levels of power had accrued to the least stable systems. Presumably, the less stable a system is, the more work is required to move a communicative vehicle from one state to another.

THE EFFICIENCY OF THE CLIENT/THERAPIST SYSTEM

As with our entropy scores, it proved impossible to interpret the results we were getting without considering the whole system. We wanted to get a sense of each system's overall efficiency. So we established a ratio between the system's final entropy score and its accumulated work score. The idea here is that the more work done to the Information Particle to increase its variety, the greater the increase in the system's communicative disorder (entropy), which, in turn, decreases the system's capacity to do work. Thus, the higher the ratio between entropy and work, the less efficient a system would be.

When we examined the results, we found the expected reciprocal relationship between a speaker's final entropy values and his or her total work values. As one increased, the other decreased, and vice versa. The thermodynamic law for physical systems so far held true in a communicative system as well.

The implication is that each Client/Therapist system has a different amount of available energy as the session begins, perhaps characteristic of the particular system. The work and entropy calculations manifest how that energy is used to change the system. Specifically, we found that the Client 2/Therapist D system (the client who championed Overeaters Anonymous and the unpredictable therapist), the most chaotic on the surface, was the most efficient in terms of work—for both client and therapist. The least efficient, in that it had the greatest loss of work capacity over the hour, was the most stable system—Client 1/Therapist B (the woman who had been married five times and the relatively silent therapist).

These findings supported results we already had. Client 2 had achieved the lowest entropy scores in her session with Therapist D, and she and Therapist D had achieved the highest accumulated work scores. Client 1 had achieved the highest entropy scores in her session with Therapist B, and Therapist B had achieved the lowest accumulated work scores.

We decided to see how the individual parties in all our sessions had contributed to the efficiency or inefficiency of their systems. We established a second ratio—between the slopes of the individual speakers' work curves.

Given the fact that the clients' work scores were always higher than the therapists', the greater the ratio, the more work the client had done to move the IP through the session. Thus, a high ratio would indicate an efficient therapist. The communicative force field establishes itself in such a way that most of its energy is available to the client. On this measure, Therapist B and Client 1 had the highest ratio, and Therapist D and Client 2 had the lowest.

This finding cast a different light on the question of systemic inefficiency. The efficient therapist may create an inefficient system—one that supports a great deal of complexity in a client's communications. The goal of therapy, after all, is to open avenues of change, and creative change tends to increase entropy.

We speculated that Therapist B's silences, along with his efforts to interpret some of the client's narrative material, had undermined the client's usual mode of expression, implicitly pressuring her to explore new variations of narrative output. This kind of approach may, over a course of therapy, compel a client to rely less on accustomed patterns of thought, thereby facilitating psychological reconstruction along new lines.

Of course, extreme degrees of entropic change might create a state of chaotic disequilibrium, which would have a pathological effect on a client. The point is that a measurably inefficient system in the short term—one that compels a client to do most of the work—may be accomplishing something long-term that a more efficient system cannot. Indeed, we had found, in our study of cross-correlational effects, that Therapist B and Client 1 had had an unusually high impact on each other's communicative behaviors.

Short-term efficiency indicates low amounts of systemic disorder. However, a course of therapy *should* be disturbing a client's maladaptive equilibrium, along with providing some basis for change. Our results strongly suggested that the energy requirements of the therapist—the amount of work needed to move his or her IP through the session—has an effect on the kind of communicative work the client can do. How we function communicatively appears to be very much dependent on our speaking partner.

THE COMMUNICATIVE FORCE FIELD

Given our evolving theory—that one speaker's use of energy in the communicative force field affects the amount of energy available to the other—we wanted to explore the nature of the force we were dealing with. Forces of nature are either conservative or nonconservative. Gravity is a conservative force. The amount of work done by gravity when something is pulled back to its point of origin is 0. A pendulum, for example, possesses potential energy when it has climbed up from its original position. When it swings down by force of gravity, that energy is converted into kinetic energy (energy in motion). But no net work is being done in that act of returning. The action of a conservative force leaves the total energy of a system unchanged—conserved—when an object returns to its point of origin.

An electromagnetic force, on the other hand, is nonconservative. A charged particle moving along a bar magnet's elliptical field lines of force will either gain or lose net energy in returning to its starting point. The gain or loss depends on the path it takes.

Our findings had indicated that the communicative force field is always giving energy to the Information Particle. Work accumulates in a dialogue as a linear function of time. Thus, during the time it takes to change states in any fashion—to visit a new state or revisit a prior state—work is accumulating. The total energy of the system has changed. The communicative force field had to be nonconservative because positive work is done in returning to a prior state.

Any path the Information Particle took in its travels would expend a varying quantity of energy. We began to wonder about the actual paths taken. There are many possible paths of return to any prior state. Some expend less energy than others. We wondered whether individual IPs take some return paths more than others.

A MATHEMATICAL DISCOVERY

As stated, *kinetic energy* is the energy of motion. The more rapidly the IP moves, the greater the value of the kinetic energy. This state of affairs is expressed by an equation that defines kinetic energy as mass times velocity squared divided by two (K.E. = m x $V^2/2$) We had decided that the mass value for the IP is always 1. So we calculated the IP's kinetic energy by noting how fast the IP moved in a particular direction, squaring that number, then dividing it in half.

In our system of measurement, work values were also found by noting how fast the IP moved in a particular direction and squaring that number ($W = V^2$). It should be clear, in this respect, that our work values, because they were not halved, were always twice as large as our kinetic energy values. This suggested that only half the work done in a session was being used to move the IP from state to state. The question became: Where is the other half of the work being applied?

There is no conceivable way this question could have arisen apart from mathematical analysis. We had assumed that all the available energy in the system, and all the work it could generate, were expended in expressive state changes. But the equations told us incisively that this was not the case.

In a mechanical process, some of the heat generated by a cycle of operations is used to perform the work involved, and some is lost to the environment. The heat that carries the system back to its prior state to begin again is called reversible heat. It is said to be reversibly absorbed because the part of the process it drives can be reversed.

But not all heat generated by a work cycle is reversible. Some of it alters the system irrevocably. This energy is called irreversible energy. The machine has been fundamentally changed, and some of its ability to perform work is lost. Caloric entropy is a measurement of the increasing disorder of a system brought about by the loss of work energy to irreversible heat. Put simply, the more cycles of operation a mechanical process undergoes, the more reversible heat absorption takes place, and the more irreversible heat is lost.

This is why the equation for determining caloric entropy involves two values: the reversible heat absorption of the system divided by its absolute temperature over all changes that occur within a defined time period. The first part of the equation is the energy that is used for work and returns to the system. The second is the temperature that indicates how much energy is in motion.

If our conjecture was right—that half the missing energy was being absorbed by the IP as reversible heat—then we had in hand a first approximation of the numerator of the equation. What we needed now was an equivalent to absolute temperature for our data. We decided to take a closer look at the movement of the IP in its five-dimensional box and see whether we could assign increments of temperature to its movements from one state to another.

CHAPTER SIXTEEN

Poetry in Motion

All life is pattern...but we can't always see the pattern when we're part of it.

—BELVA PLAIN, *CRESCENT CITY*

THE IDEA THAT communication has a temperature may seem a little silly from an everyday perspective. We speak metaphorically about hot topics, heated exchanges, and warm words, but we don't actually ascribe heat to the things people say. Temperature is something that we know by touch.

It should be remembered, however, that the property we recognize as heat occurs because the molecules of a substance are in motion. This motion can be described entirely in terms of mechanical variables, such as velocity and acceleration. Of course, we don't actually measure the billions upon billions of molecules in a physical system to figure out its temperature. That task is impractical and beyond the capacity of most computers. Instead, we get statistical averages. That's what temperature is—a statistical measure of changes that occur due to the energy of heat present in matter.

In the branch of physical science that deals with molecules, temperature is measured in units called *kelvins*, after Lord Kelvin, who established the scale. The lowest degree on that scale is absolute zero (equivalent to -273.15 degrees Celsius and -459.67 degrees Fahrenheit), the point at which molecular motion is at a minimum. It is called absolute zero because there is theoretically no lower level of atomic or molecular activity than is observed at this temperature. There are no negative kelvins. Molecules are at minimal energy or they exceed it.

Heat, in the physical world, is one way that systems exchange energy with each other. It is a universal way in which energy is accepted from other states (chemical, electrical, magnetic, and so forth) and converted freely into any of those other forms. Faster-moving molecules are colliding with slower ones and surrendering some of their energy.

We were looking for psychic parallels to this process. By measuring the velocity and acceleration of our Information Particle as it traveled from state to state, we were attempting to develop a parallel to the average measurement of kinetic energy in a system. We were assuming energy transfers of some sort, but we weren't certain yet how they might be identified.

Like molecules, the elements that make up a communicative exchange are too numerous to isolate and measure separately. We had selected our five dimensions as a representative slice of the communicative pie, believing that their behaviors would give us a statistical average for an entire exchange. We weren't convinced that our calculations in this regard would yield true parallels insofar as the mathematical models were concerned, so we did a number of preliminary studies, paying more attention to the kinds of motion we were dealing with.

According to kinetic theory, the particles of a substance can move in several different ways. *Rotation* is one of them. Because of the way our IP kept returning to prior states, we knew that its trajectory was fundamentally rotational, but we decided to see what this orbiting motion looked like geometrically. In order to do this, we invoked a proposition called Stoke's theorem, which relates the extent of rotational activity on the boundary of a system to the amount of rotation in its interior.

We were trying to find out how much work was being done in paths with the same initial and final points (which defines an orbit). Recall that work around an orbit must be positive because the communicative force is nonconservative. That is, during the time it takes to return to a prior state, work is accumulating and the total energy of the system is changing.

SOME INITIAL RESULTS

We explored these elements separately for each speaker and then looked at our averages. We discovered that roughly half the states, once visited, were revisited from two to four times. As earlier data had suggested, therapists revisited fewer states and did so more often than clients. Clients used more new states and revisited a broader range of prior states. Given the higher accumulated work scores for the clients, we wondered whether more energy was required to invoke a new state than to return to a prior one.

Interestingly, new states were nearly always adjacent to states already visited. At the level of communication, this meant that the speakers' modes of expression did not change dramatically, even when they tried out new forms. Geometrically, the phenomenon ensured near-rotational and full rotational activity in the IP. The IP was constantly circling back to favored states and states adjacent to the ones already used.

Information theory tells us that redundancy makes a message more reliable, because the message can be easily reconstructed by a receiver. Our natural preference for it is borne out by an apparently universal tendency to repeat the underlying structure of a message, and by a decreasing use of new states as information in the system accumulates. Fresh material is continually added to an established base, but less and less frequently.

Rotational Activity

The rotational quality of the IP has many physical parallels throughout nature. A charged particle, for example, will rotate along the elliptical lines of a magnetic field, and, if nothing interferes with its path, the particle will return to its point of origin. The rotation of the particles in a substance looks much like the orbits of celestial bodies, such as the moon around the earth, the earth around the sun, and so on. We wondered what kind of rotational activity we were dealing with here. Finding the rotation and classifying it was essential to build on Boltzmann's ideas.

Doing this was no easy matter. The mathematical calculations took a great deal of time. We decided to reduce the IP to two dimensions (Narrative Imagery and New Themes) and to base our next step on the results we got with them. Ultimately, we found that all of our two-dimensional IPs, whether for client or therapist, were rotating along an elliptical path.

So we engaged in the task of working out the equations for the complete IP, and the results were the same. In each of the six consultations, the IPs of both speakers rotated vigorously, constrained by a shape bounded by two concentric ellipses. One might picture a hollow watermelon, the particle running freely like a marble along the inside of the shell's rind.

Its rotation was not unlike the elliptical trajectories of planets propelled by the force field of gravity. Our trajectories were less smooth; the orbits typically zigzagged as the IP made small leaps into states adjacent to the ones already used. And there were individual differences. For example, the angles of the ellipses' axes varied from case to case, never entirely parallel to the walls of the five-dimensional box. We theorized that those variations could be analyzed for more information about a speaker's communicative tendencies. Nevertheless, the orbits were smooth enough on average to suggest that the immaterial realm of human communication shares genuine parallels with the material.

We also found that the IP virtually never visits the inner core of the ellipsoid—its central hollow—even when it's moving from low- to high-scoring quadrants. It almost seems to be a kind of forbidden zone. Abrupt shifts from non-narrative intellectualizations to highly charged stories are extremely rare. Perhaps this is an indication that nature abhors sudden reversals or shifts in direction that would take a particle from one pole to its opposite. The geometry of human communication would appear to speak for gentle, nondisruptive change.

Psychobiological Implications

Confirming some of our earlier findings, the ellipsoids encompassed no more than two to four of the thirty-two quadrants of available expression. We also found, as before, that most of the occupied space was low-score space—that is, non-narrative space. Our natural communicative state appears to be intellectual, without unconscious meaning. As suggested, this may be the state that requires the least energy to sustain.

Although all IPs periodically made sojourns into high-scoring, narrative territories, none remained there for long, and they always returned to low-score space for a more protracted stay. This tendency appears to be a major factor in the rotational activity of the IP. In fact, our data indicated that a speaker's propensity to this pattern was generally supported by the listener. Therapists tended to interrupt a client who moved toward narrative expression, either by intervening or commenting in ways that encouraged a return to strictly conscious messages.

We had seen this tendency before—in the fact that many of the therapists' interventions moved a client out of rather than into narrative communication. And we had confirmed the phenomenon mathematically. Although it was not true across the board—Therapist B was an exception—there was usually a drop in a client's new theme and narrative scores after most therapeutic comments.

In all, the patterns revealed an apparently universal tendency to rotate in and out of narrative and nonnarrative forms of communication in the course of a dialogue or monologue. This rotational activity occurs at different rates for different speakers, but it appears to be a psychobiological property of the human mind. From a psychodynamic perspective, it would suggest rotation in and out of communication capable of carrying unconscious meaning.

The fact that most of our speakers preferred nonnarrative to narrative communication suggests, as stated earlier, an inclination to limit communication that carries unconscious meaning. This may also be a fundamental property of the mind. Telling stories would appear to be a form of psychobiological adaptation—because they allow us to express highly charged unconscious perceptions—but doing so expends emotional energy. Returning quickly to the nonnarrative state may be energically necessary. Nevertheless, the shift into narrative expression clearly fuels human creativity. The invocation of narrative in our dialogues was invariably associated with invention, variation, and imagination.

Thermodynamic Implications

It isn't possible to talk about narrative meaning without engaging in subjective inference. We can also talk about the IP's rotational activity in terms of thermodynamics—strictly as a process that uses and transfers energy.

The elliptical motion we were seeing was constrained to low-score, nonnarrative space. The constraint was powerful enough that the angle at which an IP left a low-score position and moved into the domain of narratives was considerably greater than average. Thus, the journey required more than average force to get under way. One might speculate that energy has built up in an individual psychic system to the point where it must be discharged in narrative expression.

The discharge, in this respect, behaves like a Poisson event. As described earlier, Poisson events are characterized by negative exponential curves. The longer one of two states obtains, the more likely it becomes that the other will occur. With respect to our data, we would infer a tension that increases the longer a speaker remains in nonnarrative space. Once the energy has been expended in narratives, the IP quickly rotates back from high-score space into nonnarrative regions. Whatever the psychodynamic consequences of resisting unconscious communication until discharge is required, the process appears to be dictated by nature.

THE POISSON NATURE OF THE PROCESS

We decided to see if the events we were seeing did indeed conform to a Poisson model mathematically. Poisson processes were described at some length in Chapter Six, when we were dealing with the tension associated with the need to speak. A Poisson process characterizes a situation in which there are two possible states: On or Off. Let's say we're looking for a four-leaf clover (the On state). How long will we have to stand in a field picking three-leaf clovers (the Off state) before we get the next four-leaf clover we're looking for?

When we plot the actual waiting times over a series of attempts, we get a curve that is negatively exponential. That is, it shows a sequence of waiting periods (from getting three-leaf clovers to finding a next four-leaf clover) that decrease in frequency as they increase in length. The idea is that the waiting period or the Off state can obtain only so long until the On state must occur.

In terms of our data, we figured that the longer a speaker

stayed in nonnarrative space (Off), the more likely tension would build until he or she inevitably moved into narrative space (On). That is, the longer a stay in nonnarrative space lasted, the more likely the subject would move into narrative space the next second.

We configured the histograms strictly on the basis of one dimension (amount of narrative) and then spent some time generating the rate constants. The larger the rate constant, the sharper the descent or fall of the curve. As it turned out, all twelve curves were negatively exponential—not only for movement out of nonnarrative space, but for movement back into it.

In other words, the transitions from narrative to nonnarrative modes of expression are both Poisson processes, apparently reflecting a built-in human need to alternate regularly between unemotional and emotional forms of communication. For all cases, however, the narrative rate constants were larger than the nonnarrative ones, meaning that all of our speakers stayed longer in nonnarrative space than in narrative domains.

There were also individual differences. Client 1 showed some inclination to stay longer in narrative expression when she was interviewed by Therapist B, and Client 2 remained longer in both modes in her consultation with Therapist E. There was no evidence at all, however, for long rest periods of any sort. Communication—at least in these analytically oriented consultations—was an ever-cycling process, driven by a psychic force of great intensity. The speakers went in and out of nonnarrative space with a clear preference to return to it, evidently to avoid the high energy expenditure of the narrative mode.

CHAPTER SEVENTEEN

Caloric Entropy

The pure and simple truth is rarely pure and never simple.

—OSCAR WILDE

OUR INTEREST IN the geometry of the IP, as stated, was ultimately related to questions of temperature. Because temperature is essentially a measure of molecular motion in a substance, we believed we could establish a parallel to heat in a communicative vehicle by calculating the IP's velocity and acceleration from configuration to configuration. We wanted to use the resulting values to measure the *caloric entropy* of our data—that is, the decreasing capacity of the Client/Therapist system to use energy effectively.

WHAT IS CALORIC ENTROPY?

As discussed earlier, caloric entropy is the original measure of how well a system uses energy to perform work. Its value depends on how much heat the system absorbed, its temperature, and the difference between its initial and final states. The meaning of the measure is the same as the entropy calculated by Boltzmann's equation—ultimately adapted by Claude Shannon to systems of signals. However, the means of calculation differs. The Boltzmann approach is somewhat abstract—it involves an idea of information or a state. Caloric entropy is measured using intuitive and ordinary measures that associate with work, such as heat.

The larger the value of caloric entropy, the less efficient the system is in converting heat energy into work. We saw a direct parallel in the Client/Therapist system, where the idea of work is related to achieving communication between two parties. By calculating caloric entropy in the C/T system, we believed we'd discover properties of communicative interaction otherwise unrecognized.

SHIFTING MODELS

We had already analyzed our data for *informational entropy* (using Shannon's equation), which is a measure of complexity in a pattern of signals. Physics theorizes that caloric entropy is equivalent to informational entropy, which works on the basis of probabilities. To determine the growth of informational entropy in our data, we had calculated the accumulating possibilities of recurrence each time an Information Particle moved to a new configuration in the course of a dialogue.

Now we believed we had enough evidence to warrant the measurement of our data in strictly caloric terms. We wanted to construct a bridge between the material systems to which caloric entropy is normally applied and the immaterial system suggested by our findings. This is fitting because the idea of information is already an abstraction coming from a material system, such as telephone lines (or client and therapist). Ultimately, we were beginning to understand the mind as one aspect of a cosmic circle, formed by mind, matter, and energy.

In order to apply the caloric equation to our data, we treated our Information Particle as a molecule in a closed system. As two speakers interact with each other in the closed system of the Client/Therapist dyad, they create an energy pool—the *communicative force field*—and then absorb or impart energy to this field. This process is reflected in the movements (or lack of movement) of their IPs—the measurable output of their exchange. All the essential ingredients to find caloric entropy are there in the IP model.

THE BASIC IDEA

A diesel engine or a mechanical device normally performs its work by going through a cycle of operations. It goes through a number of steps, then comes back to its original state to begin again. Some of the heat generated by its operation is used to perform the work involved, and some is lost to the environment. The heat that carries the system to prior states is called reversible heat. It is said to be reversibly absorbed because the part of the process it drives can be undone or reversed.

But some of the heat in a cycle of operation is absorbed by the machine irreversibly. The heat transfer alters the system itself, and the machine's cycle can't return the system to its original values in every respect. Once heat has been transferred irreversibly, the machine or system changes fundamentally and some of its ability to perform work is lost.

Each time a machine goes through its operational paces, some of the energy it absorbs is reversible. But some energy is also absorbed irreversibly. Thus, in the course of normal operation, the absorption of irreversible energy gradually takes its toll on the system. The machine wears out. The system can do less and less work with a fixed amount of energy. Caloric entropy is a measurement of the extent to which this irreversible part of the process has taken place.

Finding a Parallel to Reversible Heat Absorption in Our Data

The equation that determines the value of caloric entropy essentially states that entropy grows as the amount of reversible heat absorption increases and/or as the absolute temperature of the system is lowered. The final value is computed by dividing reversible heat absorption by absolute temperature and summing across all changes in the system.

Although this equation is usually applied to a mechanical or physical system, it does not refer to matter as such. The only factors involved are the system's history with respect to reversible heat absorption and absolute temperature. Thus, it is possible, as Shannon ultimately demonstrated, using Boltzmann's framework, to transfer the theory to a system viewed as information.

In so far as we were concerned, the communicative system was made up of expressive vehicles—the changing states of the two IPs. As speakers rotated from prior states, to new states, to prior states, they were completing something equivalent to an operational cycle. Thus, we assumed that a return to a prior state was a direct parallel to the reversible part of a physical process. Each time the IP went through the cycle and returned to a prior state, it absorbed a certain amount of energy from the force field to support the move. Because the work done featured a return, some of that energy had to be reversibly absorbed.

Our calculations had shown, however, that only half the energy available to the IP was being used to support a state change. So we figured that this was the reversible energy—the energy absorbed and, for the most part, returned to the system in visiting some prior state. The other half, we theorized, was being retained by the IP as irreversible energy, which gradually altered the system. This formulation struck us as a little too pat, but it was a starting point.

Finding a Parallel to Absolute Temperature in Our Data

The question of *absolute temperature* was another matter. Absolute temperature is measured on the Kelvin scale and is constructed to be independent of the substance to which it applies. It is called absolute because its 0 point corresponds to the least active state of matter. Thus, we felt confident in using it to measure the internal activity of a message, as represented by our Information Particle.

As I said earlier, we were attempting to get at the issue of temperature by treating the Information Particle as though it were a kind of molecule. The faster it moved from state to state, the "hotter" the system in some sense. But temperature is really more than just a measure of motion.

The rate at which molecules give up and absorb energy is not the same for all substances and all situations. Along with several other factors, the amount of heat energy in an object depends on the mass of the object in question. We took the mass of the Information Particle to be 1, because there was no reason to assign different masses (so far) to different speakers.

Because of the expanding or growing rotation of the Information Particle, we specified two forms of motion: progressing toward new states—new configurations of the communicative vehicle—and returning to old ones. Both warranted an assignment of temperature, but the activity each one generated differed. Entry into a new state added an option of return to the system, whereas entry into a prior state did not increase the system's degrees of freedom. Accordingly, we decided the temperature assignment for new state entries should be higher than for old.

We worked out a specific set of rules for this premise—that new states predict increasing systemic activity, whereas old states predict less of an increase—and defined our units of temperature on that basis. We then wrote a program for calculating temperature on a second-by-second basis throughout a dialogue.

We tested our premises by plugging our computed values into the equation for caloric entropy and running the data. In so doing, we used our Shannon complexity results as a standard. We wanted our measurements of caloric entropy to come within 2 to 5 percent of the ones we had for complexity, and we expected them to follow the same logarithmic curves.

For this study, we used all the dialogues for which we had IP trajectory scores—seventeen in all. This made for thirty-four separate trajectories of data to run.

RUNNING THE DATA

As we had before, we chose our first case at random—Client 2 and Therapist D. This was the interview between the client who championed Overeaters Anonymous and the therapist who elicited strong responses by being unpredictable.

When we ran the values we had calculated—the sum of reversible heat absorption divided by absolute temperature—for the client, the resulting trajectory of caloric entropy formed the same kind of logarithmic curve we had gotten with informational complexity. The therapist's trajectory did the same.

Indeed, for all thirty-four trajectories, the real data fit the ideal curves even better than they had with our complexity measures. The entropy of the Information Particle was increasing quite smoothly as a logarithmic function of time.

The results were not really satisfying, however. When we compared the caloric results with our complexity results, the discrepancy was about 10 percent. We had wanted a margin of error under 5 percent. Moreover, the ranking from high to low scores (for the slopes of entropy) differed for the speakers in our two sets of entropy scores. We had missed something somewhere.

BACK TO THE DRAWING BOARD

We felt sure that our complexity measures were accurate. Shannon's equation had worked very well with our data. The problem had to be in the way we had formulated our caloric equation. Its form was more conjectural than the Shannon form.

So we decided to look beyond the paradigms we'd been using to calculate temperature. We considered the IP's parallels to a two-particle gas system. Like other kinds of systems, the absolute temperature of a gas depends on the average kinetic energy of its molecules. The value is calculated, however, as half the mass times average velocity squared. Perhaps we needed to compute the IP's velocity differently.

Velocity is speed plus direction—a two-vector measurement. We were talking, however, about velocity in a five-vector system. When we calculate the kinetic energy of a gas, which occupies three dimensions, we multiply the velocity by itself. To put this simply, we square the number because the gas has two directional possibilities in a three-dimensional space.

In a five-dimensional space, an object would hypothetically have four directional possibilities. So we tried our new equation two different ways. We tried it first the way it's constructed for gases, and saw no real improvement in our results. But when we raised velocity to the fourth power, all our speakers fell into the same order they had occupied in our complexity study.

As we had done with our final complexity scores and our accumulated work scores, we plotted absolute temperature against time across the dialogue. To our surprise, we found another perfectly linear relationship. In all seventeen dialogues, and even in our monologues, the temperature of the IP increased at a constant rate throughout the time period. The only individual differences were reflected in the slopes of our curves. In a naive sense, this is not surprising, because one can easily think of temperature as a rough measure of work. And we already knew that work grew linearly in our C/T system.

NEW PERSPECTIVES

Although we were now getting better results, and we were excited by some of the findings, we still had problems. When compared with our complexity results, the margin of error remained high. Worse, the model didn't work with every case. The inconsistencies told us that we needed to refine our measurements further.

We had been assuming, as stated, that a system's temperature increases each time a new state is visited. With a new option in play, the potential activity of the system is raised. We decided now to calculate this temperature increase more precisely—by factoring in the average number of visits per state the IP had actually made. The new method improved our findings, which led us to think further about the significance of choosing old or new states.

Therapist B, the relatively silent analyst, was a case in point. Unlike the other therapists in the study, he had achieved an unusually high ratio of new state visits to old. His client's final complexity scores had also been higher than any of the other speakers in the study. Was there a connection between the two factors that our caloric model could explain? On the basis of this question, we gradually recognized what we had been missing in our calculations. We had to account for the energy involved when the IP does *not* rotate.

The force field's rotational push is constant and powerful. It drives communicative expression back to prior states while also driving it on to new states. If a speaker does not "go with the flow"—that is, if a speaker is silent or maintains the same expressive state—he or she is doing the equivalent of swimming against the tide. The resistance requires an expenditure of energy, which increases the amount of free energy in the field.

Such resistance seems analogous to friction. The amount of energy expended would depend on the state or condition of the object being moved—that is, on the number of prior state options visited. Thus, a therapist who was more silent than active, and used more new states than old, would surrender a great deal of energy to the field.

Until this point, we had been trying to determine the amount of reversible heat absorbed at a state change on the basis of the *speaker's* average return to prior states across the session. We saw now that this was the wrong direction. The energy available to a speaker depended on how much the other speaker was taking from the communicative force field. We had to consider the *other speaker's* likelihood of returning to prior states.

Once we made the necessary adjustments to our values, the model worked with every case. Our results came to within one percent of our complexity findings. And now that we understood the principle, we could see very clearly what was happening in each of our therapeutic sessions.

The more prior states a therapist visited—that is, the more he or she returned to previously used communicative configurations—the more irreversible energy he or she absorbed, decreasing the energy available for the client's communicative work.

- The highly active therapists, who "went with the flow" of the rotational field, accumulating new states and returning to old, were invariably paired with clients whose expressive variability was low.

- The less active therapists, who resisted the flow of the rotational field, either with silence or by returning to few prior states, gave up more energy to the field. Their clients had the highest complexity scores.

TURNING TO SYSTEMIC CONSIDERATIONS

The idea that we had to include the other party's return rate in the speaking party's equation would never have occurred to us apart from the difficulties of making the math work in our models. And the success of our revised equations told us a great deal. A therapist and client are not just cooperating to create a system; each is an inherent part of the experience of the other.

What we'd been doing until now was the equivalent of attending to each note of a song improvised by two musicians. We had scored the notes separately for each—in terms of tone, pitch, and so forth—and we were now assigning increases of temperature to movements of the notes from chord to chord. But, ultimately, notes are embedded in the evolving pattern of a song. Their behavior as discrete entities has no meaning without the unifying field of the musical piece.

In the same way, the behaviors of the speakers' IPs were inexplicable without the unifying concept of the communicative field. The data had pushed us irretrievably beyond the "separate entity" Newtonian framework that had enabled our initial formulations. We were moving into a field theory of the kind championed many years ago by James Clerk Maxwell, which has influenced a great deal of research and thinking during the last generation. One speaker's communicative behaviors could not be interpreted without knowing the behaviors of the other.

We hadn't yet determined the nature of the force field we were delineating, and the variables we'd selected for measurement were not the only possible quantifiable dimensions of a conversation. Our primary purpose at this point was to pursue evidence that the laws of nature do characterize the mental as well as the physical domain.

CHAPTER EIGHTEEN

The Sounds of Silence

Drawing on my fine command of language,
I said nothing.

—ROBERT BENCHLEY

ALTHOUGH OUR FORMULATIONS had become rather complex, the central postulate remained the same: The variations of energy in the communicative force field are linked to changes of expressive state. Rotational state changes—from new configurations to old and back again—manifest the power of the force field. The IP absorbs energy from the field as it changes states and, in large measure, gives it back again.

Nonstate changes—silence, maintenance of the same configuration—are a form of resistance. They create communicative friction, and the IP loses energy to the field without absorbing any in the process. The amount of energy expended in resisting depends on how many states have been accumulated when the resistance occurs and the IP's probability of returning to them.

Clearly, "going with the flow" is the most natural course of events. It is also the most creative. Only the therapist's role as a reactive listener would suggest that a restricted form of response may be warranted.

Indeed, the idea that something as subtle as expressive state plays a critical role in the process of therapy raises a peculiar question. Are therapeutic benefits based largely on patterns of communication? What about issues of meaning and understanding?

It's possible, of course, that meaning has its greatest impact in the context of expressive patterns. However, it should be recalled that the trajectory of an Information Particle is no more than a communication's underlying shape. I spoke earlier about a bicycler, whose journey can be quantified and turned into numbers. The resulting trajectory may tell us something about the mechanics of motion and gravity, but it won't tell us anything about the pleasures of cycling on a bright summer's day.

In the same way, the trajectory of a communicative exchange may tell us something about the mechanics of psychic energy—how the mind converts it into expressive content, how people exchange it with each other in conversation, how it's distributed in different kinds of communications. Such knowledge is no small thing.

Questions about its practical application, however—interventional technique, what to listen for, what to say—involve a different process of research. For now, we can only illustrate some of the implications of our findings by referring broadly to our six core consultations. These illustrations may suggest avenues of potential application. But they cannot address issues of effective therapy without a good deal of further study.

THE MATTER OF IRREVERSIBLE ENERGY

Once we were taking into account the likelihood of both speakers returning to prior states in our calculations, it was clear that most of the heat absorbed in a state change (about 90 to 95 percent) is reversible. By "going with the flow" of the communicative field's rotational pull, speakers generate kinetic energy, use it to move their IPs from one state to another, then return most of the energy to the field to support future activity.

What we wanted to know was how much *irreversible energy* was being retained by each speaker—energy that could be used for psychic change. We found this value by taking a speaker's accumulated work total and subtracting two elements: (1) the amount of reversible heat energy generated during the session and (2) the amount of heat dissipated into the system during moments of resistance.

In light of our earlier results, the findings were predictable. The clients who had been with Therapists A and D (the most active of the six) retained the lowest amounts of irreversible energy at the end of the sessions. Therapist A, the talkative female analyst, ultimately retained more irreversible energy than her client.

The highest amount of irreversible energy was retained by Client 1 in her session with Therapist B, the relatively silent therapist who had returned to so few prior states. Among the therapists, Therapist B retained the least irreversible energy.

These findings were consistent with our newly formed ideas about silence and maintaining the same expressive state—forms of resistance to the rotational pull of the communicative force field. A therapist who intervenes intermittently, but consistently uses new forms of expression, appears to afford a client the greatest possible levels of irreversible heat energy. Highly active therapists, who return to many prior states, absorb a great deal of energy from the communicative force field, leaving little for the client to retain (presumably to effect inner change).

THERAPIST A AND CLIENT 1

Our speaker duration study had indicated that Therapist A assumed the role of primary speaker in her consultation with Client 1. She had a strong interpretive agenda, which led her to repeat the same statements to the client over and over again—for example, "The little child in you is sad"; "The mother inside you is angry"; "Something inside you is trying to get free."

Although content is not at issue here, it is apparent from these phrases why the therapist's communicative dimension scores reflect a constant return to the same general patterns of expression. She rotated into and out of a selected group of favored states—intellectualized, thematically consistent, and negative in tone.

It should be emphasized that these repeated configurations did not occur in succession, which would have indicated some resistance to the force field's rotational pull. The therapist went "with the flow," rotating between favored prior states and new states adjacent to them. As a result, the therapist accumulated a large amount of irreversible energy, and the client was compelled to sponsor it by continually absorbing and surrendering reversible heat to the field.

Significantly, the only influence this therapist showed in our cross-correlation study was limited to her repeated intellectual images. Each time she rotated back to her favorite states, the client evidently adapted her communicative behaviors to the pattern, taking little energy for herself.

THERAPIST D AND CLIENT 2

Therapist D, the unpredictable therapist, in his session with

Client 2, illustrates another kind of energy-absorbing behavior. Therapist D used a great many new states, but he also returned to a broad range of prior states. In our measures of cumulative complexity, he was the only therapist to achieve higher complexity scores than his client. Given his high rate of return, his constant variation of new states ultimately raised the temperature of his part of the system and increased its disorder, making less and less energy available to the client.

As stated earlier, this particular session had displayed only minor influence in our cross-correlation studies, all of it in the realm of positive imagery. We had wondered at the time whether the client had sealed herself off from the therapist's emotional assaults. Now it was clear that, at the level of deeper nature, she was energically isolated, obliged to hold on to whatever energy she had, rather than expend it in communicative work. This is why her complexity measures were so low.

THERAPIST B AND CLIENT 1

This session provides a good contrast to the one with Therapist A. The same client retained two very different levels of irreversible energy in these two consultations. Therapist B was silent much of the time, and the client recalled having to do too much of the "work" in this interview. When the therapist did intervene, he tended to focus on the narrative images used by the client. This is why his scores display a large amount of new states and few returns. As an apparent result of his technique, the client varied her expressive states to an unusual degree, and the communicative behaviors of both parties showed an extensive degree of interactional influence. Because the therapist did not "go with the flow" of the rotational field, his own level of irreversible energy was low, whereas the client's was highest among the twelve speakers in the study.

IMPLICATIONS OF THERAPIST DOMINANCE

As stated earlier, the therapist's role in the client/therapist system evidently ensures his or her communicative dominance. The client unconsciously adopts whatever responsive behavior will sponsor the therapist's energy needs. This hypothesis is borne out by two factors: (1) The clients' irreversible energy totals varied radically from session to session; (2) The therapists who conducted interviews with other clients (in the full consultation study) had consistent irreversible energy totals from session to session.

An intriguing question is whether the deeper energic needs of speakers play a role in the way they communicate with others. Do highly active therapists unwittingly use the therapeutic relationship to satisfy their energic needs? Is high energy absorption in a therapist's IP a possible marker for countertransference? We suspect that it is, but the issue needs further study, as does the question of client benefit and psychotherapeutic cure.

For example, we have hypothesized that Therapist B's stability and resistance to the rotation of the field allowed his client to venture beyond her usual style of communication into a form of supported disequilibrium. As suggested earlier, this is the sort of thing that might have neurological implications, leading to psychological reorganization.

It's entirely possible, however, that a therapist who enters many new expressive states makes for creative disequilibrium in another way, destabilizing the system and requiring the client to adapt in fresh ways. Conversely, a therapist who keeps returning to favored prior states may be allowing the client to "borrow" his or her communicative structure, giving a client with weak psychological boundaries a stronger sense of self.

The issues involved lie far beyond the present results. It seems clear, however, that the surface image of eclectic therapy—doing whatever works, as long as it's ethical—which has been characteristic of modern thinking about psychotherapy, is at odds with the reality of deeper nature. Therapeutic interactions are exacting; they obey the same laws of nature that obtain in the material realm. This gives us a starting point for formal research into how therapy can function in an effective way.

Indeed, the models to date enable predictions to be made based on the parameters of the therapist. If a therapist were trained to intervene such that certain parameters described his or her behaviors, then the response parameters of the client could be predicted. In other words, the present methods are evolved enough to test hypotheses about the impact of therapeutic style (as quantified by these models).

CHAPTER NINETEEN

Rethinking the Nature of the Mind

We perceive and are affected by changes too subtle to be described.

—HENRY DAVID THOREAU

SO FAR WE had concentrated most of our attention on communication between therapists and clients. We had done this for several reasons.

1. We both have a strong interest in narrative communication and unconscious meaning, particularly as it occurs in a therapy session.

2. The dimensions of such communication lent themselves to empirical description.

3. We had ten recorded consultations available for transcription and scoring.

4. The differing role expectations for therapist and client gave us reference points when we weren't certain yet how to interpret our data.

The nature of our results, however, led us to wonder more about ordinary dialogues. Would the regularities we had discovered in the therapeutic exchange appear in any kind of communicative situation: The text of a letter? A chat room on the Internet? A phone call? Was there a such thing as the entropy of small talk?

We had already been using our three nontherapy exchanges as a control study, but now we decided to consider them in their own right. The three pairs consisted of two female friends, a male and female who lived together, and marital partners. Our three monologues had been recorded by the two live-in partners and the married woman.

It was clear from our earlier studies that the couples' dialogues were comparable to the therapy sessions in several respects: They were highly stable for speaker switching and duration, showed varying degrees of interactional influence, and the curves formed by the complexity trajectories were smooth and logarithmic.

They were distinguished, however, by near-equal values for measures of final work and final entropy (both Shannon and caloric). This finding supported our ideas about the therapist determining the dominant pattern of energy absorption in a therapy session. Speakers outside of therapy would appear to absorb and provide energy in a more egalitarian fashion. In all three dialogues, the parties switched roles throughout the interaction—one speaker would talk about a problem, the other would listen and offer advice, and then their positions would reverse. The most powerful was one in which both parties allowed for alternating periods of silence and extensive use of narratives.

THE THREE MONOLOGUES

More intriguing was the task of measuring caloric entropy in our three monologues. We had been determining values on this item by factoring in the speaking partner's history of prior state visits. How would a soliloquy be conceptualized in these terms? Given our assumptions about the communicative force field, we decided to treat the monologues as interactions in which the other party's Information Particle is always in the same state. In this way, all the energy in the system is available to the speaker.

When we tried this model, we got results quite similar to the ones we'd seen with the dialogues. The laws of entropy, work, and temperature held up very well, as did the elliptical orbit of the IP. But, again, the distinguishing feature was significant.

The individuals were far more energy-efficient alone than they had been in dialogue with another person. They visited about 25 percent more states in the self-communication than they had with a partner. Without the other person's energy needs to take into account, they retained high levels of irreversible heat energy, which promoted high levels of state alteration.

Recall that the monologues had also shown high levels of cyclical power at various frequencies, a result that was absent in all of the psychotherapy data in our core study. (The only sessions in which cycles were apparent were the three extra ones conducted by Therapist B and a colleague.)

This finding may seem odd. Communication is by definition an interactive event. Yet, talking to another person would appear to distort its inherent rhythms. The apparent discrepancy struck us as significant. It indicates that the communicative force field is psychobiological in nature. It lends support to our hypothesis that communicative output is a process of energy conversion reflecting neurological activity in the brain.

Although we don't usually think about work, force, heat,

and energy when it comes to the human mind and communication, it should not be surprising that nature's myriad creations, mental and physical, share fundamental properties in common. The brain itself generates an electromagnetic field, whose measurement science takes for granted. Why shouldn't communication, one of the brain's measurable outputs, have a lawful relationship to that field?

Medical research already indicates that living neural tissue and brain synapses become physically altered in response to certain kinds of stimuli. Such alterations are called *engrams*, and they have been posited as an explanation for memory. We believe that the powerful rotational activity of the Information Particle represents the communicative equivalent of neural pathways—the pull of memory and experience toward prior expressive states. The more states visited and revisited, the deeper and stronger the engram created.

One might consider the fact that this kind of rotational activity precedes the development of language as such. Infants, whose communication with the world initially revolves around movement and sensory stimulation, will insist on hearing the same lullaby every night while beginning to explore new things. As a psychobiological property, the field's rotational character would necessarily be reflected in an individual speaker's communicative outputs more clearly when another speaker's energic needs are not involved—or not paramount.

We can see this particularly well in the consultation sessions, in which role expectations led the therapists, in general, to resist the rotational flow of the field more than their clients did. This resistance, as stated, required work, and the energy expended became available to the client. It has been pointed out many times that the Client/Therapist relationship has similarities to the mother/child system. The ideal communicative relationship for growth, learning, and development is rich in potential, but secure in structure. Reading the same story or singing the same lullaby every night is equivalent in many ways to maintaining the same expressive state over time in a dialogue. The client is implicitly encouraged to explore new states of communication while the other maintains a stable framework or boundaries.

This is especially clear in Therapist B's therapeutic behavior. His resistance to the rotational power of the field was so high, and his accumulation of states so "weighty" that he ultimately provided the energic conditions for virtual monologues from the client. Accordingly, this client used more new states than any other speaker in all ten consultations. In the extra sessions conducted by the same therapist and a colleague, the clients' trajectories behaved exactly like the ones we were seeing in our three self-communications. This suggests, again, that speaking partners sort out energy needs as a dialogue evolves, thus impeding the energy patterns that obtain when a speaker is alone.

EXTENDING THE RESEARCH TO INDIVIDUAL WORD USAGE

Whether a person speaks alone or speaks to another, the progression of communicative vehicles is always the same. The speakers move from predominantly nonnarrative expressive states to an increasing use of narrative (encoded) configurations. This cyclic activity—from narrative to nonnarrative states—is part of what drives the rotational action of the force field. As speakers absorb increasing levels of irreversible energy by changing states, there is an apparent need for discharge via work, and the speaker is compelled to move into states that involve unconscious emotional expression.

Ironically, this process is most clear in the trajectory of Therapist D, who accumulated a high level of irreversible energy as his IP rotated from new to prior states, and told more stories than any other therapist in the study. Client 1, in her session with Therapist B, accumulated the highest degree of irreversible energy, and her IP spent more time in narrative space than anyone else among our thirty-four speakers.

We decided to pursue this issue in a way that would also tell us more about the nature of nontherapeutic communication. We developed a study that would focus on one of the simplest aspects of human communication—how people select individual words.

Our analysis could proceed on the basis of completely objective data, because no scoring would be required. We simply programmed a computer to replace each word in a text with a position number and to enter that number each time the word was used again.

For example:

A	=	1
rose	=	2
by	=	3
any	=	4
other	=	5
name	=	6
is	=	7
yet	=	8
a	=	1
rose	=	2

With this coding, the emergence of new words corresponds to a new integer and a repeat of a prior word to the reoccurrence of a former integer. We wanted to know whether the selection of new words as a text progressed had any similarities to the selection of new states in a dialogue.

Complexity and Word Usage—The First Sample

We decided to begin with creative texts and used a sample of poems from seven well-known poets: Lord Byron (*She Walks in Beauty*), William Shakespeare ("Shall I compare thee to a Summer's day?"—Sonnet 18), William Ernest Henley (*Invictus*), William Wordsworth (*The World Is Too Much with Us*), Edgar Allan Poe (*Annabel Lee*), Robert Frost (*The Road Not Taken*), and Samuel Taylor Coleridge (*Kubla Khan*).

We also included, for comparison's sake, a sample of prose by Lewis Carroll from the beginning of Chapter I of *Alice's Adventures in Wonderland.*

To explore these new data, we turned to two measures that had proven themselves in our earlier studies. The first was the growth of Shannon entropy—or complexity. This would tell us about the amount of accumulated variety in the use and reuse of words in a given sample.

The second was a model similar to the one we had used to study speaker duration and the waiting times between utterances from one speaker or the other. We applied this model to the waiting times (specifically the number of words that occur) between the appearances of new words. In this way, we could determine any regularities to the process of using new words. Was there any counterpart in a text sample to the tension that built in a dialogue when one person spoke for a long time?

Results of the Study

In every sample we used, the cumulative entropy of individual word selection increased logarithmically, just as it had in our dialogues.

There were individual differences, of course. For one thing, the poems were of varied lengths, which made the possibilities for maximum complexity different from poem to poem. And the rate at which complexity increased was also distinctive for each writer. Shakespeare showed the greatest use of available complexity, and Poe showed the least. But these differences operated within the constraints of an apparently universal law.

The bar graphs, or histograms, for the waiting times between new words again took the form of negative exponential curves—the telltale fingerprint of a Poisson process. This suggests that the longer an individual continues to return to previously used words, the greater the buildup of tension to make the next word a new one. Evidently, many aspects of the use of language by the human mind are characterized by tension-building mechanisms that ensure variety in the midst of stability.

Here, too, there were individual tendencies. Shakespeare and Poe again formed the two ends of the spectrum. Shakespeare used many new words, so that the waiting times between them were low. Poe's waiting times were high.

The Follow-Up Study

We were excited by these findings and decided to run other kinds of writing—essays, scientific material, famous orations by Abraham Lincoln and Martin Luther King, Jr. The results were remarkably alike. Poets in general showed more complex configurations of word usage than other kinds of writers and speakers, but all our graphs showed the same logarithmic curve.

The tendencies we were documenting seemed so universal and so fundamental to human functioning that we extended our research to a study to the communicative output of chimpanzees. We obtained samples of chimpanzee communication from animals who had learned to use language by pointing to symbols on a board of lexigrams as they interacted with their trainers.

Despite the small vocabularies of the primates and the slowness with which they communicate, this data showed the same regularities we had seen in human word selection—the increasing buildup of complexity and the negative exponential curves with respect to waiting times between new "words." Indeed, their curves were graceful and complex, resembling the results we had obtained from our poets. Perhaps the absence of cultural inhibitors had allowed the chimpanzees to express themselves more freely than most humans!

ASSESSING THE NEW RESULTS

We have introduced these results for two reasons. First, they suggest that our findings with dimensions of communicative expression involve a fundamental property of the mind. What we found to be true at the higher level of geometry and process dynamics holds true at the more primitive level of word usage. Even as we reuse words, we are building a tension that is not relieved until the moment we vary our selection. The shifting use of new and prior states, as well as the shifting use of narrative and nonnarrative expression, would appear to be different aspects of the same phenomenon. Nature tends to repeat form and structure at many levels of the cosmic scale.

This is why the fractal concept is so enchanting. A fractal is a geometric pattern that is repeated at ever smaller scales to produce irregular shapes and surfaces that cannot be represented by classical geometry. The same form persists in appearance at all levels of magnification. The fractal here lies in the models that manifest across all levels of communication, including primates.

The invocation of variety and instability on the one hand, and sameness and stability on the other, appears to be an adaptive mechanism. It may well characterize any mind, human or primate, capable of representative expression—whether by sign, signal, symbol, metaphor, or encoded image. One would have to speculate a psychobiological origin for the tendency, rooted in the brain and its neuronal activities.

CHAPTER TWENTY

The Cosmic Circle

Truth is stranger than fiction, but it is because Fiction is obliged to stick to possibilities; Truth isn't.

—MARK TWAIN

IN THE FIRST chapter of this book, I spoke about the peculiar nature of most psychological research, which sets itself the paradoxical task of measuring the immeasurable—thoughts, feelings, dreams, wishes: all the immaterial stuff the subjective mind is made of. This paradox has been the stumbling block to a formal science of the mind; it has also perpetuated the philosophical conundrum of the mind's relationship to the brain. The brain is clearly a material entity with measurable properties, but the mind is understood as immaterial, undefinable in terms of such things as energy requirements, chemical and electrical activities, and anatomy. We end up with the classic idea of the ghost in the machine.

The paradox, however, is a false one. After all, we measure the brain's functioning by way of its outputs in electrical signals and chemical products. We have attempted in our research to show that the mind can be measured in the same terms—by way of its communications: the observable representations of thoughts, feelings, dreams, and wishes. Just as we learned through painstaking experimentation which aspects of the brain are critical to accurate description and measurement, we can set up experiments to identify and quantify critical attributes of the mind. Our first tentative attempts at such experimentation have indicated that the laws of the mind and the laws of the brain are isomorphic or even identical. The same mathematical models can be used to describe the behaviors of both realms.

Our findings have forced us to the conclusion that the brain and the mind are each a particular expression of matter and energy. Each has its own domain, mode of expression, and manner of adhering to nature's laws. And even though the mind appears to arise from the brain, each entity can be treated separately—as two aspects of a single system.

In this respect, identity theory, which proposes that the mind and brain are expressions of the same fundamental entity, and dualism, which holds that they are separate and interacting entities, are both true.

EXPLORING THE MIND/BRAIN RELATIONSHIP

The brain does appear to be the source of energy for both systems, drawing sustenance and power from chemicals in the blood supply and perhaps also from the Sun's light. We would propose that communication is a conversion process, transforming the energies of the brain into a communicative force field that fuels mental outputs.

Energy transformations of this kind are common throughout nature. Shake a conducting wire and the kinetic energy of motion creates a magnetic field, whose properties are distinctive to itself. One might speculate that units of mental energy, like the units of magnetic energy, are measurable at a fundamental level. The Information Particle, in this respect, may represent some aspect of a larger entity that has real, physical properties.

There is nothing mystical about this proposition. If material energies are the fueling source for mental energies, the existence of the latter would depend on whatever physical supplies the material substrate requires. The human brain, after all, consumes about 20 percent of the body's available energy resources. We have no reason, on the basis of our data, to hypothesize nonmaterial sources of mental energy, although we have no reason, on the same basis, to rule them out. The possibility simply isn't relevant to the results we obtained.

Once the existence of mental energies is established and accepted, mind, matter and energy become unified into a single system within nature. Conversion of one into another takes place within that grand system, indicating that there is indeed a complete, deep unity within the universe.

This proposition—of a cosmic circle—does have mystical implications, but our myths, faith traditions, and philosophies are often containers for truths that science hasn't yet appropriated. It hearkens back to the classic idea of prime matter and allows the primal stuff to manifest as matter, energy, or mind. Matter becomes superdense energy, and mind becomes superdense form to accommodate matter, and so on. Mind, like energy, would be massless yet convertible to matter.

MIND, MATTER, AND ENERGY

In Genesis, the creation of the cosmos is a cataclysmic process of separation—heaven from earth, light from dark, land from water. Particle physicists tell us that the forces of nature were indeed once unified into a single force, which broke apart as the universe evolved. Their separation generated four identifiable forces—the strong and weak forces, electromagnetism, and gravity. Evolution, from this perspective, has largely been a process of these forces interacting with different forms of energy. One localized form of energy is matter, whose properties bind space and time to our ordinary experience of life.

As recent arrivals to this evolutionary drama, humans have sought to locate their place and meaning in the unfolding process. The consciousness that drives this search has suggested to more than one philosopher that Mind was also present at the beginning of time, another aspect of the matter and energy continuum.

Accordingly, we suggest that Mind has always been involved in the ordered creation and direction of form. We certainly have evidence for Mind in other physical entities—for example in the communication of subatomic particles in zero time and their apparent ability to make informed choices. One might say, in this broad sense, that Mind is present wherever form and regularity exist. It is Mind that uses force to exploit the instrumentality of energy, thus shaping the myriad forms of material life. The cosmic circle closes when matter returns to its point of origin and realizes its portion in Mind.

One might also say that matter embodies all of its potential forms within its confines. The form of observed matter at any given point in time reflects but one of its realizable forms. In this respect, the spiritual traditions are right when they accord a form of consciousness to rocks and trees and organic life of all kinds.

The Information Particle, in this sense, belongs to that arc of the cosmic circle which connects Mind to energy in terms of the communicative force field. But the IP is also subject to the laws of physics, which connects energy to matter.

Communication, understood in this light, is a way of watching the universe unfold. There are few faith traditions on earth that do not celebrate the advent of representational thought as sacred, particularly as it issues in language and art. Communication is the confluence of the three states of nature, making possible their recognition of identity in each other. Their differences are in their states, not their contents. Quite remarkably, in the human brain all three transformations (mind into energy, energy into matter, and matter into mind) take place.

What we are saying is not unlike the propositions of some faith traditions regarding the material universe as the gods' way of contacting each other and becoming aware of themselves. Mind, in this respect, is not equivalent to God, but one of the media through which God acts and is manifest. To participate in that activity is to recognize meaning in the unfolding of the universe. To put this more plainly, we are the universe looking at itself, a now familiar idea.

ACROSS THE CUTTING EDGE

Such speculations bring us, full circle, back to areas touched on in the beginning of the book. The field of parapsychology, as suggested, has had the same problems with measurement as the field of psychology. Both fields, however, depend on observable signs that suggest the existence of an immaterial entity. The research we've outlined in this book might lend itself to areas of extrasensory communication as well as to sensory communication. A psychic dream, for example, is narrated; its meaning depends on the use of language.

Moreover, our research has shown scientifically that, without knowing it, individuals create certain kinds of coordinated rhythmic dances in their dialogues. They have deep and powerful effects on each other that can be faithfully documented, although never experienced or represented consciously or unconsciously. And we have fashioned a reasonable hypothesis that there are deep laws of the mind and a communicative or informational force field that is activated by human dialogues and interactions.

Indeed, this same force field may well be activated by

human monologues, chimpanzees attempting to communicate with humans, and perhaps, by any effort at sign or symbol communication. There is a broad range of potential implications of research into these forces and laws.

We might suggest, for example, a study in which two people—a sender and a receiver—are asked to engage in a narrative exercise at the same moments in time. One could then measure the extent to which the deeper patterns in these dialogues are coherent. This is an aspect of dialogue that has no conscious counterpart; it cannot be "fudged" or anticipated. If coherency was found between the sender and receiver, and was absent between control subjects, we'd have a remarkable piece of evidence for extrasensory communication.

Our cross-correlation measures could serve to assess the same effects. Jung's concept of synchronicity between the physical and psychic worlds might also be testable with the methods we've outlined.

PARTING THOUGHTS

We have allowed ourselves the liberty of presenting a mixture of mathematically defined results, reasonable conjectures, and sheer speculation. We have now reached an esoteric area of phenomena that stands, at the moment, beyond science. Ultimately, however, this entire book was about dreams.

We dreamed of creating a sound foundation for a science of the mind; of characterizing reliable therapeutic measures; of contributing to a new understanding of mind, matter, and energy. With the help of science and the force of imagination, we may yet see these dreams come to pass.

BOOKS AND SCIENTIFIC PAPERS

Background references can be found in the following publications:

1. Langs, R. (1987). Psychoanalysis as an Aristotelian science: Pathways to Copernicus and a modern-day approach. *Contemporary Psychoanalysis,* 23: 555–576.
2. Langs, R. (1988). Perspectives on psychoanalysis as a late arrival to the family of sciences. *Contemporary Psychoanalysis,* 24: 397–419.
3. Langs R. (1988). Mathematics for psychoanalysis. *British Journal of Psychotherapy,* 5: 204–212.
4. Langs, R. (1989). A systems theory for psychoanalysis. *Contemporary Psychoanalysis,* 25: 371–392.
5. Langs, R. (1989). Models, theory, and research strategies: Toward the evolution of new paradigms. *Psychoanalytic Inquiry,* 9: 305–331.
6. Langs, R. J., & Badalamenti, A. F. (1990). Stochastic analysis of the duration of the speaker role in psychotherapy. *Perceptual and Motor Skills,* 70: 675–689.
7. Badalamenti, A., & Langs, R. (1990). An empirical investigation of human dyadic systems in the time and frequency domains. *Behavioral Science,* 36: 100–114.
8. Langs, R., & Badalamenti, A. (1990). Quantitative studies of the therapeutic interaction guided by consideration of unconscious communication. In *Psychotherapy Research: An International Review of Programmatic Studies.* Ed., Beutler, L. E., & Crago, M. Washington, D.C.: American Psychological Association, pp. 294–298.
9. Langs, R. J., & Badalamenti, A. F. (1991). Statistics and mathematics in psychotherapy research. *Bulletin of the Society for Psychoanalytic Psychotherapy,* 6: 13–21.
10. Langs, R. J., Badalamenti, A., & Bryant, R. (1991). A measure of linear influence between patient and therapist. *Psychological Reports,* 69: 355–368.
11. Rapp, P.; Jimenez-Montano, M.; Langs, R.; Thomson, L.; & Mees, A. (1991). Toward a quantitative characterization of patient-therapist communication. *Mathematical Biosciences,* 105: 207–227.
12. Badalamenti, A., & Langs, R. (1992). Stochastic analysis of the duration of the speaker role in the psychotherapy of an AIDS patient. *American Journal of Psychotherapy,* 46: 207–225.

13. Badalamenti, A.; Langs, R.; & Kessler, M. (1992). Stochastic progression of new states in psychotherapy. *Statistics in Medicine.* 11: 231–242.
14. Badalamenti, A., & Langs, R. (1992). Stochastic analysis of the duration of the speaker role in the psychotherapy of an AIDS patient. *American Journal of Psychotherapy,* 46: 207–225.
15. Langs, R., & Badalamenti, A. (1992). The three modes of the science of psychoanalysis. *American Journal of Psychotherapy,* 46: 163–182.
16. Langs, R.; Badalamenti, A.; & Cramer, G. (1992). The formal mode of the science of psychoanalysis: Studies of two patient/therapist systems. *American Journal of Psychotherapy,* 46: 226–239.
17. Langs, R.; Rapp, P.; Thomson, L.; Pinto, A.; Cramer, G.; & Badalamenti, A. (1992). Three quantitative studies of gender and identity in psychotherapy consultations. *American Journal of Psychotherapy,* 46: 183–206.
18. Langs, R. (1992). Toward building psychoanalytically based mathematical models of psychotherapeutic paradigms. In *Analysis of Dynamic Psychological Systems, Vol. 2.* Ed., Levine, R., & Fitzgerald, H. New York: Plenum Press.
19. Badalamenti, A., & Langs, R. (1992). The thermodynamics of psychotherapeutic communication. *Behavioral Science,* 37: 157–180.
20. Badalamenti, A., & Langs, R. (1992). Work and force in psychotherapy. *Journal of Mathematical and Computer Modeling,* 16: 3–17.
21. Badalamenti, A., & Langs, R. (1992). The progression of entropy of a five-dimensional psychotherapeutic system. *Systems Research,* 9: 3–28.
22. Langs, R., & Badalamenti, A. (1992). Some clinical consequences of a formal science for psychoanalysis and psychotherapy. *American Journal of Psychotherapy,* 46: 611–619.
23. Langs, R. (1992). *Science, Systems, and Psychoanalysis.* London: Karnac Books.
24. Langs, R. J.; Udoff, A.; Bucci, W.; Cramer, G.; & Thomson, L. (1993). Two methods of assessing unconscious communication in psychotherapy. *Psychoanalytic Psychology,* 10: 1–13.
25. Badalamenti, A.; Langs, R.; & Robinson, J. (1993). Lawful systems dynamics in how poets chose their words. *Behavioral Science,* 39: 46–71.
26. Langs, R. (1993). Psychoanalysis: Narrative myth or narrative science? *Contemporary Psychoanalysis,* 29: 555–594.
27. Badalamenti, A.; Langs, R.; & Cramer, G. (1993). The non-random

nature of communication in psychotherapy. *Systems Research,* 10: 25–39.

28. Langs, R., & Badalamenti, A. (1994). Psychotherapy: The search for chaos, the discovery of determinism. *Australian and New Zealand Journal of Psychiatry,* 28: 68–81.
29. Badalamenti, A., & Langs, R. (1994). A formal science for psychoanalysis. *British Journal of Psychotherapy,* 11: 92–104.
30. Badalamenti, A.; Langs, R.; Cramer, G.; & Robinson, J. (1994). Poisson evolution in word selection. *Mathematical and Computer Modelling,* 19: 27–36.
31. Langs, R. (1995). *Clinical Practice and the Architecture of the Mind.* London: Karnac Books.
32. Langs, R. (1995). Psychoanalysis and the science of evolution. *American Journal of Psychotherapy,* 49: 47–58.
33. Langs, R. (1996). Mental Darwinism and the evolution of the hominid mind. *American Journal of Psychotherapy,* 50: 103–124.
34. Langs, R.; Badalamenti, A.; & Savage-Rumbaugh, S. (1996). Two mathematically defined expressive language structures in humans and chimpanzees. *Behavioral Science,* 41: 124–135.
35. Langs, R. (1996). *The Evolution of the Emotion Processing Mind, With an Introduction to Mental Darwinism.* London: Karnac Books.

Index

adaptation, 8–9, 79

boundaries, 7–8, 15–16, 79

chimpanzees, word usage of, 196
client/therapist (C/T):
  system, 39–43, 134–137, 158–160, 172–173, 180
  unity, 160, 179–180
clinical (therapy) material, 25–34, 59–71, 80, 82, 85, 87, 88–89, 92–93, 95, 97–98, 99–100
coherence, harmonic, 112, 116–117
communication, ix–xi
  markers, 53–55
  unconscious (encoded), 9–15
communicative vehicle (five dimensions), x–xi, 18, 57–58, 72–73, 76–78, 79–81, 126–127
  selection of (state selection), 138–139, 165–167, 174–176, 178–180, 181–185
correlations (statistics), 38, 72–73, 76–78, 81–89, 92–105,141
cosmic circle, 173, 201–203

emotions, ix–x, 109
influence, deep, 75–78, 81–89
information particle (IP), 126, 138–139, 148–151, 165–169, 173–176, 177–178, 180, 181, 202

language (word usage), ix, xi, 9, 16, 193–197
mathematics, 20–24, 38–39, 40
mind:
  and brain, 199–201
  and matter, energy, 200–203
narratives (stories), ix–x, 9–17, 104, 137–138, 167–170
nonnarrative communication, 167–170
nontherapy (everyday):
  dialogues, 52–53, 103, 117, 190,
  monologues, 117, 191–193

parapsychology, 5–6, 204
physics, models based on:
  electricity (PSD function), 109–118, 141–142
  entropy:
    informational, 119–125, 127–129, 131–137, 141–143, 172–173, 194–196
    caloric, 143, 171–180, 183
    law of, emotional, 131–134
  power, cyclical, 113–116
  temperature, 163–168
  work and force, 146–152, 155–162
    force field, communicative, 147–148, 152, 160–161, 173,

178–180, 181–185, 203–204
energy of, 160–162, 174, 178–180, 183–185, 200–203
Poisson process, 47–50, 55, 101, 169–170, 195–196
psychoanalysis (psychotherapy), 1–6, 21–22, 25–36, 37–38, 71–72,
science of, x–xi, 1–6, 19–24, 37–38, 145, 182

speaker duration:
stability and instability of, 41–44, 51–52, 77, 101–102
study of, 39–44, 46–55
stochastic (probabilistic) models, 40–44

therapist dominance, 54–55, 186–187
thermodynamics, 119, 122–123, 141, 168–169

unconscious mind (experience), x, 4–6, 7–16